T0215259

CAMBRIDGE LIBRARY COLLECTION

Books of enduring scholarly value

Botany and Horticulture

Until the nineteenth century, the investigation of natural phenomena, plants and animals was considered either the preserve of elite scholars or a pastime for the leisured upper classes. As increasing academic rigour and systematisation was brought to the study of 'natural history', its subdisciplines were adopted into university curricula, and learned societies (such as the Royal Horticultural Society, founded in 1804) were established to support research in these areas. A related development was strong enthusiasm for exotic garden plants, which resulted in plant collecting expeditions to every corner of the globe, sometimes with tragic consequences. This series includes accounts of some of those expeditions, detailed reference works on the flora of different regions, and practical advice for amateur and professional gardeners.

Selected Papers, 1821–38

Professor of botany from 1825 until his death, John Stevens Henslow (1796–1861) revived and greatly advanced the study of plants at Cambridge. His influence helped to make the University Botanic Garden an important centre for teaching and research. Originally published over seventeen years, and now reissued here together, these thirteen papers reveal the impressive breadth of Henslow's scientific knowledge. The first two items, from 1821, address the geology of the Isle of Man and Anglesey respectively, preceding his five-year tenure of the chair of mineralogy at Cambridge from 1822. The rest of the papers, dating from 1829 to 1838, address botanical topics. John S. Parker, Director of Cambridge University Botanic Garden (1996–2010), has provided a foreword that traces Henslow's developing interests and contextualises the items in this collection. Several of Henslow's other publications, including his *Catalogue of British Plants* (1829), are reissued separately in this series.

Cambridge University Press has long been a pioneer in the reissuing of out-of-print titles from its own backlist, producing digital reprints of books that are still sought after by scholars and students but could not be reprinted economically using traditional technology. The Cambridge Library Collection extends this activity to a wider range of books which are still of importance to researchers and professionals, either for the source material they contain, or as landmarks in the history of their academic discipline.

Drawing from the world-renowned collections in the Cambridge University Library and other partner libraries, and guided by the advice of experts in each subject area, Cambridge University Press is using state-of-the-art scanning machines in its own Printing House to capture the content of each book selected for inclusion. The files are processed to give a consistently clear, crisp image, and the books finished to the high quality standard for which the Press is recognised around the world. The latest print-on-demand technology ensures that the books will remain available indefinitely, and that orders for single or multiple copies can quickly be supplied.

The Cambridge Library Collection brings back to life books of enduring scholarly value (including out-of-copyright works originally issued by other publishers) across a wide range of disciplines in the humanities and social sciences and in science and technology.

Selected Papers, 1821–38

John Stevens Henslow

CAMBRIDGE
UNIVERSITY PRESS

CAMBRIDGE
UNIVERSITY PRESS

University Printing House, Cambridge, CB2 8BS, United Kingdom

Published in the United States of America by Cambridge University Press, New York

Cambridge University Press is part of the University of Cambridge.
It furthers the University's mission by disseminating knowledge in the pursuit of
education, learning and research at the highest international levels of excellence.

www.cambridge.org
Information on this title: www.cambridge.org/9781108070546

© in this compilation Cambridge University Press 2014

This edition first published 1821–38
This digitally printed version 2014

ISBN 978-1-108-07054-6 Paperback

Foreword: The Science of John Stevens Henslow
John S. Parker
Director, Cambridge University Botanic Garden (1996–2010)

In the received history of modern evolutionary theory, Cambridge University's Professor John Stevens Henslow (1796–1861) has been awarded a small but highly significant role. He is remembered as the man who recommended Charles Darwin to be the companion of Captain Robert FitzRoy and act as a naturalist during the voyage of H.M.S. *Beagle*. Henslow wrote to Darwin in August 1831 indicating that he regarded Darwin as being 'amply qualified for collecting, observing and noting anything worthy to be noted in natural history', although, with a fine appreciation of his own role as teacher and mentor, he gave the admonishment that Darwin was not yet a 'finished naturalist'.

In his letter to Leonard Jenyns, included in the *Memoir* that Jenyns published after Henslow's death, Darwin says of Henslow that 'his accurate powers of observation, sound sense and cautious judgement seem predominant ... drawing conclusions from minute observations ... [he also] shows his capacity for extended observations and broad views' (Jenyns, 1862). Darwin, however, prefers at this time, soon after Henslow's death, to remember his lifelong friend above all for his gentle, courteous demeanour, rather than emphasising the quality of his intellect and his contribution to science. Darwin's great friend Sir Joseph Hooker, the eminent botanist and Director of the Royal Botanic Gardens, Kew, from his perspective as Henslow's son-in-law, gives his own poignant view of Henslow's life. Hooker's eulogy of his father-in-law records the passing of a great scientist, but one whose character outshone even his scientific brilliance. His death 'is like gouging a piece out of the face of the country ... I miss his knowledge [of] Botany and loads of kindred subjects'. More, he was one of 'those friends formed ... to be a lamp unto our path whom we never go ahead of' (Huxley, 1918).

The conventional interpretation of Henslow's influence on Darwin's intellectual development at Cambridge and on his subsequent life is partial, and indeed misleading in its brevity and simplicity. In Henslow, Darwin found a guide to the natural world who expanded his horizons from that of a collector with an obsession for beetles to a brilliant and insightful field naturalist versed in geology, zoology and botany, capable of making profound judgements about his environment from observation alone. This volume in the Cambridge Library Collection presents a selection of Henslow's papers over a seventeen-year period, which together demonstrate the breadth of his scientific research and thinking, and which qualified him to be regarded by Darwin as someone 'of whom we were all awe-struck with the amount of his knowledge ... he knew everything'.

Early Life of John Stevens Henslow: 1796–1823

Henslow was born in Rochester, Kent, in 1796. His paternal grandfather, also John Henslow, had moved from the south coast to Somerset House in London when he became Chief Surveyor of the Navy, and was later knighted for his energetic commissioning of naval ships which subsequently saw service in the war with France. He married a Kentish woman when based at the Royal Naval Dockyard at Chatham. Their son, John Prentis Henslow, was initially a lawyer in Enfield, and then joined a relative as a wine merchant in Rochester (Walters and Stow, 2001).

From his riverside home at Frindsbury, on the banks of the River Medway opposite Rochester, young John Stevens would have seen on each side the swelling uplift of the North Downs with their chalky fields and hangers rich in a diversity of plants, animals and fossils. The family recorded how he was an ardent collector from his earliest years, staggering back from country walks under loads of curious stones and fascinating sticks which took his eye. He also had an early talent for drawing and painting, which he employed throughout his career for recording his scientific observations and in teaching.

Henslow's maternal grandfather was Thomas Stevens, a wealthy brewer in Rochester, and John was named for him. Thomas built a house called Gads Hill, high among the rolling downs in a wooded area a few miles to the south-west of Rochester. Henslow visited his grandfather frequently, even when he was a Cambridge professor with a wife and children, finding this biologically rich countryside very appealing and intriguing. He discovered and collected here many rare plants and animals, which were very different from those found in the river valley of the Medway Gap and the coastal salt marshes below the hills.

After a few years' education in Rochester, Henslow was sent in 1805 to Wilson's Grammar School at Camberwell, then a village in rolling countryside to the south of London, to continue his education up to university entrance under the headmastership of Rev. W. Jephson. Here he was particularly influenced by the art master George Samuel, whose hobby was entomology. Henslow learnt the techniques of art from him, through drawing and painting the butterflies and other insects which they collected around the village. Samuel was friendly with the noted young zoologist William Elford Leach, appointed Assistant Curator of the British Museum in 1813, and the famous entomologist James Stephens, and the enthusiastic schoolboy Henslow was introduced to these two influential men by him. Later, while at Cambridge, he helped Leach at the British Museum during vacations and submitted specimens, particularly of marine organisms, for incorporation into the British Museum collections. Henslow developed into a knowledgeable entomologist, becoming one of Stephens's friends and carrying on a long correspondence with him about the British fauna.

Rev. Jephson had been educated at St John's College, Cambridge, where his brother was a fellow, and so it was natural for Henslow to be admitted there as an undergraduate in October 1814, aged eighteen. He opted to read mathematics, and was regularly among the top prizemen at the college in each year of study. He graduated in 1818 as a wrangler, placed sixteenth in the University list. He would perhaps have finished even higher in mathematics if he had not devoted a lot of his energy to enthusiastically absorbing a whole spectrum of the natural sciences during his undergraduate career. For example, Henslow developed an interest in molluscs and devised a net for extracting tiny shells from the muddy bottoms of the rivers of the Fens, an early example of specific ecological instrumentation. One of the bivalves he collected was new to science, and was named for him by Leonard Jenyns as *Pisidium henslowanum*. He presented one of his marine specimens from Devon, collected in 1817, to the British Museum and it was also named after him: 'Henslow's swimming crab', *Polybius henslowii*. Proposed by his friends William Leach and James Stephens, he was elected a fellow of the Linnean Society of London for his entomological and other zoological knowledge in February 1818, a month after he graduated in mathematics.

One of the great attractions at Cambridge in the first two decades of the century was the lecture course of Edward Clarke of Jesus College, for whom the Chair of Mineralogy was created in 1808. Although their science content was rather slight, each lecture was a tour de force owing to his flamboyant style and eclectic experiences gained on a trip round Europe, the Levant and Egypt (Otter, 1824). The lectures attracted many attendees, including Henslow, who absorbed from them knowledge of mineralogy, and more importantly crystallography.

Henslow was also introduced, when an undergraduate, to Adam Sedgwick, who was ten years older and already a fellow of Trinity College. Together they developed their interest in the young science of geology, and both joined the hub of this new science, the Geological Society of London. In 1818, Sedgwick became Professor of Geology at Cambridge, despite confessing his relative ignorance of the science, and acknowledging the equal expertise of his young friend Henslow, who attended Sedgwick's first course of public lectures given in Lent Term of 1819. After this the pair departed in the Easter vacation for a geological tour of the Isle of Wight to gain the field experience which both men lacked (Jenyns, 1862).

After their walking tour exploring the geological diversity of the Isle of Wight, Henslow now felt confident in his abilities in field geology and immediately followed this up in the summer of 1819 by leading a party of undergraduates on an exploration of the Isle of Man. Here his prime role was to give instruction to the students in preparation for their examination, but he also collected plants and studied the geology of the island. His geological results did not agree with interpretations given in a previous geological survey, so he prepared a paper, including beautifully drawn sketches and maps, which was published in the *Transactions of the Geological Society of London* in 1821. This was Henslow's first scientific paper (see the first item in this volume).

A very different outcome of the Isle of Wight tour was a proposal by the two friends to promote the cause of science at Cambridge through a society for discussion and debate on scientific matters within the university. With support from Professor Clarke, Sedgwick and Henslow spread this idea among the dons, and many eminent scholars supported them (Hall, 1969). After a public meeting in 1819, the Cambridge Philosophical Society was formally set up in 1821; it flourishes still. Henslow and Sedgwick were founding signators, and Henslow acted as secretary from its inception until 1839, when he resigned on his move from Cambridge to a rural parish at Hitcham in Suffolk.

On his graduation in 1818, Henslow had not performed well enough to be made a fellow of St John's College. Since his father, John Prentis Henslow, was rather unsuccessful in both the law and as a wine merchant, John had no independent income and so had to find ways to support himself if he wished to remain in Cambridge. One way to raise funds was to tutor undergraduates for examinations as he had done on his Isle of Man trip, but he found another and became demonstrator to the Professor of Chemistry, James Cumming. Cumming was particularly interested in electrochemistry, on the borders of what we would now regard as chemistry and physics, so Henslow expanded his scientific knowledge to encompass these sciences too.

Henslow's job was to carry out experiments at the front of the lecture theatre under Cumming's direction, for the benefit of the undergraduate audience. Later, one of Henslow's revolutions in Cambridge teaching was to get students personally involved in practical science. His experience as a demonstrator clearly showed him the unsatisfactory nature of practical science training at the time – observing others rather than through personal investigation. He addressed this deficit when he was able.

Henslow's interest in zoology continued to grow, and it becomes clear at this time that his fascination was not taxonomy but processes – physiology and anatomy allied to morphology. For example, he made minutely detailed drawings of his dissections of snails, in which he linked the organs he displayed to the recent terminology introduced by French anatomists. Unfortunately, he left this early work unpublished (see drawings in the Bath Royal Literary and Scientific Institution archive).

In the summer of 1820, Henslow furthered his interest in the geology of islands with a visit to Anglesey. Again, he was accompanied by undergraduates whom he instructed in the Cambridge examination syllabus. His prime objective, however, was to make a geological survey and to map the strata of the island. He collected about a thousand rock specimens. Back in Cambridge he annotated and arranged them for Sedgwick and the developing geological museum. By studying his contributions to the University Herbarium, we know that he also spent a part of most days on this trip in collecting plants.

By late 1821, his study of Anglesey was complete, and he read his long and detailed paper to meetings of the Philosophical Society. It was then published in the first volume of the *Transactions of the Cambridge Philosophical Society* and is a massive work of nearly one hundred pages, accompanied by a superb colour map summarising his findings and many sketches (see the second item in this volume). It is clear that Darwin's own attempt at a major geological field analysis of the Falkland Islands in 1834 was modelled exactly on Henslow's paper on Anglesey. From the detailed commentary in his notes, Darwin must have had the paper with him on the *Beagle* voyage (Darwin, n.d.). In his letter to Jenyns for the *Memoir* in 1862, Darwin singles out Henslow's 'admirable memoir on the geology of Anglesea' for particular praise.

Although most of the early records of Henslow's biological enthusiasms concern animals, it is clear that he gave his attention to botany too. His oldest surviving herbarium specimens were collected in 1816. One of these was the rare British herb from which woad is obtained, *Isatis tinctoria*. Another is a herbarium sheet displaying 'monstrous' (fasciated) forms of *Plantago major*, illustrating this developmental abnormality. The phenomenon of monstrosity fascinated Henslow and was the subject of several later papers. The unusual nature of these sheets suggests perhaps that Henslow collected many more herbarium specimens but disposed of the commoner ones when the systematic preparation of a British herbarium became one of his top priorities from 1821 onwards. Several interesting plants also survive from his 1819 Isle of Man visit, and over one hundred from Anglesey.

In 1820, Henslow was introduced to Leonard Jenyns, an undergraduate at St John's College and one of the few people in Cambridge at that time who shared his interest in zoology. This acquaintance grew to become a lifelong friendship, and was further sealed three years later by Henslow's marriage to Jenyns' sister Harriet. In the *Memoir* published after Henslow's death, Jenyns records that in 1820 they decided together to make a collection of the whole British flora as dried specimens on herbarium sheets (Jenyns, 1862).

Henslow started his plant collection with enthusiasm in March 1821 and collected about two hundred flowering plant specimens in this first year. Jenyns was primarily a zoologist and proved only intermittently committed to their botanical project, although he frequently accompanied Henslow on collecting trips around Cambridge, and also contributed interesting plants whenever he came across them. From the outset, however, it is clear that Henslow's object in making collections was distinctive and unique – he collected not simply to have specimens of all species, as a taxonomist would, but to explore the patterns of variation in nature in order to define the species by that means. Thus his earliest herbarium sheet was of the tiny moss *Weissia lanceolata*, from the Gog Magog Hills to the south of Cambridge, collected on 29 March 1821. It displays a meticulous and beautiful arrangement of fifteen individual moss plants, in ascending order of size.

Professor of Mineralogy: 1822–7

In 1822, Professor Clarke died and the Chair of Mineralogy became vacant. Henslow applied for it and was elected, at the young age of twenty-six. He was clearly regarded in the University as the most appropriate candidate because of his wide scientific background and the esteem in which he was held both as a result of the publication of his two important geological papers and for his position as secretary of the Cambridge Philosophical Society.

Henslow immediately began the process of devising a new syllabus for his course of mineralogy lectures. These he based on completely different principles to those of his predecessor. Henslow was inspired by the work of the Abbé Haüy, the French crystallographer. Haüy had, around the turn of the nineteenth century, found a way of describing the structure of crystals by applying mathematical principles, and any newly discovered crystal could be ascribed to its place within mathematically defined families – order was produced out of chaos for this facet of the natural world. The syllabus Henslow published for his lecture course has a long introduction, in which he argues the case for the Haüy system (Henslow, 1823). By contrast, Clarke's lectures had had no philosophical basis, but consisted of a series of topics, many of them economic.

Only one research paper emerged from Henslow's tenure of the Chair of Mineralogy: 'On the Crystallization of Gold', published in the first volume of the *Magazine of Natural History* in 1828. Indeed, Henslow's scientific publication record between 1822 and 1828 is rather sparse, but he was far from idle during this time. He busied himself with teaching his mineralogy course each year, by getting married in 1823 and starting a family in 1825, by taking holy orders in 1823 and carrying out duties as curate of Little St Mary's Church in Cambridge, and – more significantly for science – by collecting and mounting in his herbarium thousands of plants which he obtained from innumerable field excursions around Cambridge (and occasionally around Rochester when visiting his family).

To increase his British herbarium still further, Henslow began to build up a massive network of collectors across the whole country. There were fellow professors, such as Hooker at Glasgow University and Graham and Balfour at Edinburgh, but he profited particularly from a huge number of knowledgeable amateurs – his own close relatives, his friends and acquaintances, as well as known botanical enthusiasts identified through the Linnean Society of London. With these amateur collaborators, he exchanged lists of his requirements, his 'desiderata', in a reciprocal way. Thus, Cambridge specialities, particularly fen plants, were incorporated into herbaria across Britain. The number of his acquisitions reached a peak in 1826, when he added more than 800 plants to his collection.

Professor of Botany: 1825 onwards

In the first two decades of the nineteenth century, botany was a moribund subject at Cambridge. The Chair of Botany had been founded in 1727 and the third holder, Thomas Martyn, was appointed in 1762. Martyn was initially very active, involving himself vigorously with the establishment and financing of a Botanic Garden in the centre of the city, giving lectures, and writing extensively about his subject (Walters, 1981). His enthusiasm, however, had waned by the 1790s and he moved permanently to his parish in Bedfordshire. He retained his chair but in a dormant state, so that for about thirty years no botanical lectures were given in the University, while the Botanic Garden began to slide into decay. He eventually died in 1825 at the age of eighty-nine, and John Henslow successfully applied for his position. So, by the age of twenty-nine, Henslow held two chairs at Cambridge, and in very different subjects. His great love, however, was botany, and after two years of holding both positions he resigned from the professorship of mineralogy. He retained the Chair of Botany until his death in 1861 and, despite failing health, he taught his summer course every year up to 1860.

As with mineralogy three years earlier, Henslow immediately began the preparation of his botanical lecture course. He spent the first three months of 1827 working intensively using his artistic skills to produce about seventy large (20x26 inches) watercolour illustrations of botanical subjects for display during his lectures. The course itself began in May 1827, and the lectures were the first in the English universities to be illustrated. William Darwin Fox was so enthusiastic about Henslow's first course that he told his cousin Charles that it was not to be missed. Darwin himself said that 'His lectures on botany were universally popular, and as clear as daylight'. So accurate and clear were Henslow's 1827 teaching diagrams that they continued to be used in the Botany School at Cambridge for undergraduate lectures until the 1960s.

Henslow's botany course was revolutionary in many ways. As well as the novelty of illustration during lectures, he provided each student with equipment such as needles, scalpels, dissecting tiles and microscopes. Each week, he gathered plants from the Botanic Garden and from nature for the students to investigate for themselves – the first practical science for all, with Henslow giving training in detailed observation of natural phenomena.

A number of excursions were timetabled during the five-week botanical course. Henslow took his students into the field to examine the different habitats around Cambridge. Lists of the plants seen were then printed and distributed after each field trip. At the same time, his friend Sedgwick was running field excursions in geology for his students, sometimes covering great sweeps of country on horseback. Henslow's progresses were rather more sedate – on foot through the fens and over the Gog Magog Hills; by barge down the Cam to Baits Bite Lock; and by coach to an unusual, rich, acid-soil heath at Gamlingay to the west of Cambridge. While attempting to collect the water plant *Utricularia* for Professor Henslow on this boggy heath at Gamlingay, Darwin slid gracefully underwater in a ditch, to the great amusement of his fellow students. Darwin, however, emerged triumphantly clutching his prize.

Henslow's Botanical Studies

Although Henslow had become an active botanist by 1821, the results of his studies emerged as publications after a long delay due to his other preoccupations. In 1827, however, his resignation from the Chair of Mineralogy enabled him to devote all his energies to his beloved plants, and *A Catalogue of British Plants* was ready for use by his students during his third annual botanical course in 1829. This list was revolutionary as it was not based on the work of Linnaeus but on the 'natural system' favoured by the French school led by A.P. de Candolle. Henslow's knowledge of species came from his own remarkable, unique British herbarium. Even today the University Herbarium at Cambridge holds 3,654 Henslow sheets carrying over 10,000 plants, and this still represents 89 per cent of the 1,200 or so species recognised by him in his *Catalogue* (Henslow, 1829).

For Henslow, the focus of his herbarium studies was not taxonomic. He was not concerned with the classification of plants in a 'methodical manner, according to some ... of various methods or systems [of classification]'. His interests lay in 'Physiological Botany, as this subject possesses more general interest, owing to the numerous and striking phenomena ... which it enables us to explain' (Henslow, 1836). Henslow was thus an experimentalist interested in how to 'connect the numerous facts ... and laws which regulate the functions performed by the living vegetable'. He viewed his collection of dried specimens as a tool to address the most important question of the day in botany – the laws that regulate the *variation of species*.

To address this species question, Henslow assembled his herbarium sheets by a process he called 'collation' (Kohn et al., 2005). A collated sheet contains several plants (a maximum of thirty-two) showing aspects of variation (height, for example) arranged in distinctive patterns. We would today recognise these patterns as representing bell curves, or ascending or descending series. Henslow the mathematician is revealed here. By compiling collated herbarium sheets using the range of specimens he had, Henslow was seeking what he called the 'limits' of each species. These results he then presented in his *Catalogue*. Remarkably, then, Henslow was arguing that species, and the variation within each species, could be subject to experimental investigation. And the stimulus for this view of species can be seen in the work of Abbé Haüy on crystallography. Henslow wrote in 1836 that, after the discovery of the laws of crystallography, 'a single crystal at once puts the mineralogist in possession of ... the species, and he can calculate "à priori" the possible forms under which it may occur' (Henslow, 1837). The act of collation was an attempt to transfer Haüy's ideas from the geological to the botanical world, and so resolve disputes about what constitutes a species.

Henslow was fully committed to the idea of the creation of immutable species at some distant time in the past, the orthodoxy among the scientists of his day. However, he developed a fine appreciation of the nature of species during the 1820s – exploring the extent and importance of variation, trying to understand the role of 'monstrosity' as a key to understanding the (developmental) laws that govern plant growth and differentiation, and establishing the importance of hybridisation as an experimental method of determining the *limits* of species. Many of Henslow's subsequent botanical papers emerge from this appreciation of species.

Importantly, Henslow's most favoured student, Charles Darwin, was instructed in these fundamentals during his innumerable interactions with him. Darwin's understanding of the importance of variation and species, of populations, and the problems presented by monstrosities, enabled him to evaluate his own observations and experiences as he circumnavigated the globe on the *Beagle*, and subsequently as he pondered them on his return. Using Henslow's approach, however, he reached radically different conclusions, and the stable species became unstable.

Most of Henslow's botanical investigations emerged from observations he made during his innumerable collecting trips into the countryside around Cambridge, often alone, with friends such as Jenyns, or later with his students. His 'experimental' approach to the species problem necessitated close observation of living plants to detect patterns of variation. Thus he noticed the fringed edge of the leaves of the tiny bog orchid *Malaxis* (now *Hammarbya*) *paludosa* (see item 3 in this volume). Microscopic observation revealed to Henslow that the minute outgrowths were vegetative propagules or gemmae, and he illustrated this with his own drawings. His pencil sketches are still held in the University Herbarium.

Henslow also realised that spatial considerations were important to an understanding of botany, and so he meticulously recorded the locations of all the plants he collected. This probably reflects his interest in the broader, world-scale patterns of plant distribution which were emerging at the time, the study of which we now call 'biogeography'. Thus his papers on 'monstrosities' of the flowers of common mignonette (1835; see item 9) and *Adoxa* (1837; see item 11) arose from chance observations of aberrant individuals he found on his rambles near Cambridge and whose localities he specifies. He clearly had an exceptional eye for developmental variants. For example, his herbarium contains two aberrant forms of *Linaria vulgaris*, the yellow toadflax, found around Cambridge. In one, the standard bilateral flower with a single nectar spur takes on a radial symmetry with five spurs (a *peloric* form), while the other lacks the spurs completely (form *ecalcarata*).

Henslow clearly developed a sense of the concept of populations, and his herbarium specimens reveal this. For example, he collected seven flowering individuals of *Orchis fusca* (now *O. purpurea*) from Boxley Wood, near his grandfather's house in Kent, on a single visit in May 1827. One is extremely tall and robust, others are small and spindly, some have many-flowered spikes and some few-flowered. He clearly mounted them on sheets to illustrate the variation occurring within this single population.

One of the most remarkable of Henslow's papers, however, concerns a single population of the rare lily-like herb *Paris quadrifolia*, which he discovered in a new and isolated site in a wood at Coton, about three miles west of Cambridge. Henslow noticed that individuals of this species differed in the numbers of their vegetative and floral parts, so over a three-year period he collected information from 1,500 plants in this population on Whitwell Farm. From each individual he recorded five numerical parameters – numbers of leaves in the sub-floral whorls, numbers of sepals, petals, stamens and styles within flowers. He also recorded and illustrated all the aberrations of stamen morphology he encountered. As far as I am aware, this is the first modern study of a biological population, and is a compilation of 7,500 bits of data. In this mammoth task, he was helped by two undergraduate students from his botany course, Charles Babington (who later succeeded him as Professor of Botany) and John Downes (who supplied Henslow with plants showing variation from near his home in Northamptonshire).

In his paper on the varieties of *Paris* in the Whitwell population (see item 7), Henslow demonstrates his mathematical background in his logical presentation, in tabular form, of this complex multivariate data. He also uses further tables to explore the *'limits'* (his italics) within which the observed variation occurs. He then considers his observations in a wider context. He considers the related genus *Trillium,* in which the floral parts are strictly in threes, and proposes from its variation that *Paris* is 'ever struggling to become double in all its parts'. Henslow's capacity for combining such detailed observation with broad speculation, as shown here, was referred to by Darwin in his eulogy in Jenyns' *Memoir* of 1862. Interestingly, one of Darwin's numerous enquires of Henslow was to ask whether populations at the edge of a range could still show variation despite their isolation and their inevitably small numbers of founders. Variation is clearly necessary for evolution to occur at the edge of a range. A manuscript in the Darwin Papers at Cambridge University Library is annotated with a word from Henslow associated with this question – *'Paris'.*

In considerations of the nature of species throughout the nineteenth century, the genus *Primula* played a highly significant role. Darwin's *The Different Forms of Flowers on Plants of the Same Species* (1877) is primarily a record of experiments he carried out to reveal the consequences of cross- and self-fertilisation, using the 'pin' and 'thrum' forms of primroses (*P. vulgaris*) and cowslips (*P. veris*). Darwin uses this breeding evidence as support for the theories on the significance of sexual reproduction that he had advanced in *On the Origin of Species* (1859). Henslow, in 1830, while Darwin was his student, debated the nature of species and the necessity for experimentation to resolve issues in his paper 'On the Specific Identity of the Primrose, Oxlip, Cowslip and Polyanthus' (see item 4). As a result of observations of populations of *Primula* he found at Westhoe, near Cambridge, combined with experiments in his own garden, Henslow began to doubt the existence of separate species in this genus, contrary to the thinking of 'modern botanists'. He concludes that the view of Linnaeus eighty years previously, that the different forms are varieties of a single species, was indeed correct.

Hybridity is clearly an important consideration for Henslow in his thinking about the definition of species, and in the *Primula* paper he calls for rigorous experimentation on hybrids in many different genera to underpin observation. He also urges botanists to tabulate their experiments, and not ignore failures and 'unsatisfactory' results. In a paper published in the same year, he uses the fact of segregation of different forms of offspring from a single capsule of *Anagallis* to refute the existence of two species based on flower colour – *A. arvensis* and *A. coerulea* – in favour of them being two varieties of the one species *A. arvensis* (see items 5). Thus hybridity and segregation together provided Henslow with a definitive test of species distinction. Indeed, he later went further and proposed that if observations on a hundred different hybrids and their parents could be collected, then the laws of heredity could be elucidated.

Henslow understood that hybrids are formed in nature as well as being produced by direct intervention. When an unusual *Digitalis* (foxglove) arose in his own garden in Cambridge, Henslow quickly realised its hybrid origin, and followed its development throughout its flowering and fruiting life since it provided him with 'curious and important physiological facts' (see item 6). His detailed description of the developmental progress of the hybrid and its two parents is masterly, and is one of the most important early contributions to this area of research, although he modestly considers he had 'thrown so little light … upon the great questions connected with the phenomenon of hybridity'. What he had done, however, was to set the stage for future considerations of the nature of species, through his influence over his students, and particularly by imbuing Charles Darwin with this unique understanding.

Darwin came to Cambridge in 1828 as a failed medical student from Edinburgh University. He was to read for an Ordinary Degree, which would qualify him for a comfortable career as a parson in the Church of England. At Cambridge, Darwin was pointed towards Professor Henslow by his cousin, William Darwin Fox, who had attended Henslow's inaugural botany course. Darwin found

Henslow inspiring, and attended the botany course in 1829, 1830 and 1831. He also took full advantage of Henslow's generous, open and novel pedagogy, notably attending Henslow's Friday evening soirées at which the great men of Cambridge assembled to discuss science. Indeed, Darwin spent so much time with his mentor that he became known to the dons as 'the man who walks with Henslow'. Henslow's ultimate act of generosity came when he recommended Darwin for the position of gentleman companion to Captain Robert FitzRoy on the circumnavigation of the globe by H.M.S. *Beagle*.

At Cambridge, Henslow received all the material collected by Darwin on the voyage, and dealt with it. Darwin made a particular point of collecting the kinds of plant material he thought would be most acceptable to Henslow, including population samples. As it turned out, Henslow the experimental botanist was inadequate to attempt a full taxonomic analysis of these plants, although he mounted beautifully more than 2,000 specimens and meticulously labelled them. However, he did describe and name two of the endemic species of *Opuntia* from the Galápagos archipelago in 1837 (see item 12), and he compiled a short flora (a *florula*) of the native plants of the Keeling Islands the following year (see item 13).

It later fell to Joseph Hooker to address the taxonomy of Darwin's Galápagos specimens and give a definitive account of the remarkable endemism shown by the plants of this remote archipelago. The botany of isolated islands was as intriguing to Henslow, who lectured on biogeography, as it was to Darwin. Henslow the scientist was at pains to urge his fellow botanists to distinguish between introduced plants and those that were indigenous to an area when compiling floras and plant lists, in order to 'arrive at a knowledge of the laws which regulate the geographical distribution of species'. In his own *Catalogue of British Plants* (1829) he printed in italics all species not so far met with in Cambridgeshire.

Henslow's views on the science of botany can be gleaned from the syllabus of his first lecture course of 1828 and its subsequent editions (Henslow, 1828). The substance of these early lectures was later brought together in book form as *The Principles of Descriptive and Physiological Botany*, published in 1836 in Dionysius Lardner's 'Cabinet Cyclopædia' series and reissued in the Cambridge Library Collection in 2009. One of the great influences on Henslow's own thinking was the work of the French-speaking botanist A.P. de Candolle. Their common thinking on modern botany and the importance of physiology (non-taxonomic botany) is evident in Henslow's 48-page review, published in the *Foreign Quarterly Review*, of de Candolle's 1832 three-volume work *Physiologie végétale* (see item 8). Henslow uses this review to consider his own philosophical position, although, curiously, he presents the review anonymously.

During the 1830s, Henslow became increasingly involved with national scientific developments and the dissemination of knowledge throughout the country and at all levels in society. He was a staunch supporter of the British Association for the Advancement of Science from 1831, organising the third of its annual meetings at Cambridge (Morrell and Thackray, 1881), and was a founder member of the Royal Agricultural Society of England (Goddard, 1988). He was also intimately involved in the rise of the Mechanics' Institute movement, which sought to spread education among working men – his membership card for the Cambridge Mechanics' Institute bears the number 2.

Alongside his increasing national position as an academic, Henslow advanced in his clerical career and received the living of Cholsey-cum-Moulsford in Berkshire in 1832. In 1837, he was offered a more prestigious, and much more financially rewarding, parish at Hitcham in rural Suffolk. After two years as an absentee rector, Henslow made the momentous decision to move his family to this rather remote agricultural parish (Russell-Gebbett, 1977). From then on, he focused his scientific endeavours on agriculture and its improvement on a rational basis, publishing papers on such subjects as wheat and its diseases, potatoes and clover, and manuring and plant nutrition. However, he maintained his Cambridge connections and returned from Suffolk every summer to give his botanical lecture course until a year before his death in 1861.

References

Darwin, C.R. (n.d.) 'Geological Diary'. Cambridge University Library, DAR33.217–222.

Goddard, N. (1988) *Harvests of Change: History of the Royal Agricultural Society of England*. Quiller Press, Fakenham.

Hall, A.R. (1969) *The Cambridge Philosophical Society: A History 1819–1969*. Cambridge Philosophical Society, Cambridge.

Henslow, J.S. (1823) *A Syllabus of a Course of Lectures on Mineralogy*. James Hodson, Cambridge. [Reissued by CUP – ISBN 9781108002011]

Henslow, J.S. (1828) *Syllabus of a Course of Botanical Lectures*. James Hodson, Cambridge.

Henslow, J.S. (1829) *A Catalogue of British Plants Arranged According to the Natural System*. James Hodson, Cambridge.

Henslow, J.S. (1836) *The Principles of Descriptive and Physiological Botany*. Longman, London. [Reissued by CUP – ISBN 9781108001861]

Henslow, J.S. (1837) 'On the Requisites Necessary for the Advance of Botany'. *Magazine of Zoology and Botany*, 1, 113–125.

Huxley, L. (1918) *Life and Letters of Sir Joseph Dalton Hooker*. John Murray, London. [Reissued by CUP – ISBN 9781108031028]

Jenyns, L. (1862) *Memoir of the Rev. John Stevens Henslow*. John van Voorst, London. [Reissued by CUP – ISBN 9781108035200]

Kohn, D., Murrell, G., Parker, J., and Whitehorn, M. (2005) 'What Henslow Taught Darwin'. *Nature*, 436, 643–645.

Morrell, J. and Thackray, A. (1981) *Gentlemen of Science: Early Years of the British Association for the Advancement of Science*. Oxford University Press, Oxford.

Otter, W. (1824) *Life and Remains of Edward Daniel Clarke*. J.F. Dove, London.

Russell-Gebbett, J. (1977) *Henslow of Hitcham: Botanist, Educationalist and Cleryman*. Terence Dalton, Lavenham.

Walters, S.M. (1981) *The Shaping of Cambridge Botany*. Cambridge University Press, Cambridge. [Reissued by CUP – ISBN 9781108002301]

Walters, S.M. and Snow, E.A. (2001) *Darwin's Mentor: John Stevens Henslow, 1796–1861*. Cambridge University Press, Cambridge. [Reissued by CUP – ISBN 9780521117999]

TRANSACTIONS

OF THE

GEOLOGICAL SOCIETY,

ESTABLISHED NOVEMBER 13, 1807.

VOLUME THE FIFTH.

Quod si cui mortalium cordi et curæ sit, non tantum inventis hærere, atque iis uti, sed ad ulteriora penetrare; atque non disputando adversarium, sed opere naturam vincere; denique non belle et probabiliter opinari, sed certo et ostensive scire; tales, tanquam veri scientiarum filii, nobis (si videbitur) se adjungant.

Novum Organum, Præfatio.

London:

PRINTED AND SOLD BY WILLIAM PHILLIPS, GEORGE-YARD, LOMBARD-STREET.

1821.

XXVII.—*Supplementary Observations to Dr. Berger's Account of the Isle of Man.*

By J. S. HENSLOW, Esq.

MEMBER OF THE GEOLOGICAL SOCIETY.

[Read April 7, 1820.]

IN laying before the Society the following remarks, I beg leave to state that my intention is not to attempt a minute description of the geological features of the Isle of Man, but chiefly to point out the localities of several formations unnoticed by Dr. Berger in his account of the Island, inserted in the second volume of the Transactions. With this view I shall not trespass on the time of the Society, by repeating the description of any place he may have already mentioned, unless where I have thought him too brief or erroneous; but, as I have sometimes found it necessary to differ from him, I must offer in my defence the information I have received from gentlemen of this Society, who have assured me, that his account was drawn up from loose memoranda, long after he had left the Island, when, probably many facts had escaped his memory. At the same time, having his observations before me, I was naturally induced to become more anxious in my research wherever I imagined there appeared to be any inaccuracy in his descriptions.

The knowledge of the few additional facts that I now offer may tend to promote further enquiry into the structure of this Island; but much yet remains to be done before a complete account can be expected.

Dr. Berger himself admits his map to be faulty, and I was entirely destitute of the means of accurately correcting it; but some of the positions are so erroneously laid down as materially to affect what I have to communicate, and I therefore submit to the Society a rough sketch of the mountains (Plate 35), particularly of the central chain, which will, I hope, serve to convey at least a general idea of their disposition. The chief error in Dr. Berger's map is in the delineation of the southern group, the whole of which has been brought three or four miles too much to the east; by which means the road from Castletown to St. John's is thrown on the west, instead of the opposite side of South Barrule. Cronknyirrea-Lhaa also, as well as the mountainous ground stretching from it to Brada head, is made to rise about midway between the east and west coasts, instead of forming an overhanging cliff to the latter.

A central chain may be distinctly traced from north Barrule to Cronknyirrea-Lhaa, bearing north-east and south-west, a direction nearly parallel to the greatest length of the island. This chain includes most of the highest mountains, the tops of which are nearly in the same straight line, with the exception of Cronknyirrea-Lhaa, which lying rather to the north of this direction, gives the southern extremity a slight curvature towards the west.

As cursory observations require no particular method of arrangement, I shall follow that adopted by Dr. Berger, beginning with granite and gneiss.

Granite exists in much greater quantity in the Isle of Man, than appears to have been suspected by Dr. Berger, and may be traced further than he supposed it to extend, even where he mentions having found it. This was near the Dun, or as the Manks pronounce it, the Dtooden, a small harbour about three miles to the north of Laxey. A stream discharges itself here, and at a short distance from its mouth receives two feeders from the north, the most easterly of which is again subdivided into two courses, and it was in the southern branch of this that Dr. Berger perceived the decomposed portion of the granite he mentions. The different courses descend from a boggy land on the hill to the west, and run across the high road, but the main stream on the south is the only one of sufficient magnitude to require a bridge. Both branches of the northern feeder run over the granite as far down as their union, where it disappears. The southern branch, at one spot, runs through a vaulted excavation in its course below the road, where the rock is so soft as to admit of its being dug by the spade. If a westerly course be pursued up the hill from this spot, many places will be found where the bare rock is exposed for several yards square, and more particularly on some eminences near the top, where the soil has been washed away, and which are situate above the bog before alluded to, near the source of a stream which crosses the road about half a mile to the north, at Ballallen. It may also be traced towards the east, up the How, and though no good section is afforded in this direction, yet the disposition of numerous large blocks, unmixed with those of any other rock, sufficiently point out its extent.

This granite passes into gneiss, which is also in a state of decomposition in the excavation above mentioned, though the small portion denuded precludes the possibility of ascertaining its connexion with the granite, and confines the description to a bare exposition of the fact of its existence.

Dr. Berger mentions the composition of this granite only at the spot where it occurs in a decomposed state; but at greater elevations it is perfectly compact, the quartz and mica bearing as large a proportion to the felspar as usual. The mica is black and the varieties of texture numerous, but it is impossible to trace with any degree of certainty the beds which constitute the different changes, as no very extensive section is formed in the rock. All that can be stated is, that the greater the elevation, the more compact and less decomposed the aspect; and on approaching the stratified rocks to the south, the mica gradually yields to chlorite, the quartz at the same time becoming less abundant, till some varieties are little else than felspar and chlorite.

If I have been prolix in my directions for detecting the locality of the granite here, it has arisen from the desire of leading future geologists to the spot without that loss of time which might otherwise attend their search. For it is situate between two streams not half a mile distant from each other, whose courses any inquirer would naturally be led to explore. On finding the dip and direction of the stratified rocks which form the bed of each to be the same, he would most probably conclude that they were parts of a continuous bed, and thus overlook the granite which occupies the intermediate space: a circumstance, which not only happened to myself upon first visiting the spot, but even, as I was informed, to a celebrated geologist, who was there the previous year.

The above is the only spot where Dr. Berger mentions having met with granite *in situ*; but another more extensive tract of this nature occurs to the south of Foxdale. The main body forms a hill or ridge, stretching nearly north and south for about one mile, and is called by the inhabitants "Slieu-ny-clough," or "the Stony Mountain." The best section of this is formed by the course of the river which runs through Foxdale, and there are also several places on its summit which are laid bare in a manner similar to those mentioned near the Dun. The stream affording this section rises on the eastern side of south Barrule, crosses the road between Castletown and St. John's near the sixth mile-stone, and after flowing through a small extent of boggy land, suddenly turns down the slope on the side of Slieu-ny-clough in a northerly direction. Another section is formed by a stream running on the same side, but down the opposite slope of the hill, and here the granite is soon lost beneath mica-slate, but towards Kirk Marown it once more appears at an elevation not so great as before.

The aspect of the granite forming Slieu-ny-clough is very different from that found near the Dun; the materials of which it is composed are in general large grained, especially towards the summit of the hill. On comparison with various granites in the collection of G. B. Greenough, Esq. it appears closely allied to that from Dalky, in Ireland, and also to some towards the southern extremity of Scotland, and it is worthy of remark, that these three localities are situate in nearly the same straight line.

On Slieu-ny-clough itself, I did not perceive any gneiss, but between this and Kirk Marown church, it occurs in a few places; at a hill about midway, behind the Garth, and close to the church. At these places it rises through the clay-slate presenting a bare rock in the midst of verdure. The felspar often greatly predominates; it then has a compact texture with a few spots of mica and quartz dispersed through it; the fracture conchoidal; the structure thick slaty. A similar gneiss is met with on Slieu-ny-carnaane. No section is formed in it, but it appears to lie in a bed penetrating the mountain in a direction about north-east by east. Broken masses occur disposed in an elevated ridge, and which, from the direction of quartz veins traversing several contiguous blocks, may be known to be still *in situ.* It will readily be conceived that the granite found in working the Foxdale mines, mentioned by Dr. Berger, page 36, was a continuation of that forming Slieu-ny-Clough, since the southernmost shaft is within gun-shot of the spot where we first meet with it denuded on ascending the course of the river.

QUARTZOSE DISTRICT.

By this title I wish to designate three localities, throughout which the prevailing ingredient of the rocks is quartz, and as I know not under what head to class them, I prefer giving a description of their appearance to hazarding a conjecture. No good section is afforded which might enable us to compare their connexion with the other formations, some quarries however occur at intervals which give us an opportunity of examining their structure, which in several places consists of quartz only, very crystalline and finely granular, and would deserve the name of quartz rock, were it not associated with other varieties. The first district to be mentioned lies between the Dun granite and Slieu-Roy, on which latter mountain many large blocks are scattered, which lie bleached on the surface, and at a distance present the appearance so commonly exhibited in granite regions. A section on the west of the mountain shews this rock intimately associated with the clay-slate in broad contorted beds, the composition here consisting of the same ingredients as granite in a very comminuted state; but the general character of the scattered blocks is that of a genuine quartz rock.

To the south-east of South Barrule and Cronk-ny-irrea Lhaa, the quartz is blended with mica and specks of clay-slate, sometimes with the latter only, presenting an intermediate passage between mica-slate and clay-slate.   On Slieu-ny-Carnaane the appearance is similar, and as Dr. Berger notices the occurrence of blocks of mica-slate on this mountain, he probably considered the rock of which it is composed to be of that nature.

MICA-SLATE.

Two distinct varieties of this are met with on Slieu-ny-Clough. The first occurs in laminæ about an inch in thickness, where the Foxdale river bends to the north after leaving the Castle-town road. The surfaces are richly coated with mica, giving the specimen a brilliant appearance when detached from the bed.   It is traversed by veins of quartz and schorl, sometimes crystallized in alternate bands.

The other variety is met with in two or three detached places lower down the stream.   The rock has a greyish aspect, and consists of fine grains of quartz and mica intimately mixed with small crystals of schorl.   Layers, one or two inches thick, of granular quartz, sometimes lie between contiguous laminæ, and inclose small reddish-brown garnets.   Other layers have larger crystals embedded of a liver-brown colour.   Although the bed of the river is sunk through this to the depth of ten or twelve feet, and we can trace both it and the granite along different portions of its bottom, the actual contact is no where visible, but is concealed by broken masses and rubbish.

CLAY-SLATE.

Dr. Berger divides this into clay-slate and greywacke-slate, and the latter does in fact appear to exist, though the gradation between it and the former is so insensible that it is impossible to draw the line of separation, and I have preferred designating the whole tract occupied by these under the name of clay-slate, the general character being that of the Cornish killas.

The gneiss near the Dun, passes into a chlorite slate very regularly stratified. The strata run north-east and south-west, dipping at a considerable angle to the south-east. A bed occurs here about twelve or fourteen feet wide, which lies between well defined strata, but is itself in a state of confusion. It appears to have once consisted of strata similar to those which enclose it, but which from some disturbing cause have been broken and bent in every direction at the time of their consolidation. A sketch (fig. 5, Plate 35.) of a portion of this ten feet in length, representing a vertical section formed by the stream at the Dun, will convey some idea of their disposition. It is much decomposed on the surface, having assumed a brown tinge to the depth of about one foot into the rock. A vein partly composed of clay, and partly of a steatitic substance united with a considerable portion of lime and a few specks of galena, traverses this. The bed may be traced from the road side a little beyond the bridge and along the course of the stream, about half way between the former spot and the sea, until it is lost there. Owing to the inclined position of the strata, connected with the rapid descent from the top of the hill to the sea shore, the real disposition of the bed has not a direction so nearly east and west, as the section formed at the surface might appear to indicate. It is from this bed lying conformably to the direction and dip of the laminæ of clay-slate, that I presume the strata and fissile texture are in this spot coincident. The clay slate on the western coast, between the northern termination of the red sandstone and Kirk Michael, does not assume that regularly slaty appearance which it presents on the opposite side of the island. There occur also in several places nodules of quartz sparingly dispersed, and near Balla Neah, I observed the cliffs to consist of angular fragments of clay-slate embedded in a clay-slate paste, and what is curious, these fragments are scarcely to be distinguished from the base excepting on the surface of the rock which has been exposed to the action of the waves, where they become sufficiently apparent by the fragments assuming different tinges of colour, giving the specimen a mottled appearance. One or two beds of magnesian limestone and iron pyrites occur here.

GREYWACKE.

Beds of greywackè occur in the clay-slate which do not lie conformably with the direction of the laminæ of the latter. The description of one will serve to convey a general idea of the others. On the north part of Douglass-bay at the top of the hill over which the road passes upon quitting the shore, is a quarry in which one of these occur. Two sections at right angles to each other afford a convenient opportunity of examining its position.

Fig. 1. Plate 35, is a section perpendicular to the direction of the strata, and fig. 2, parallel to them. The direction of the bed is nearly north and south (that of the strata being north-east and south-west,) and its dip is to the west. In one part it is seen traversing the clay-slate with great regularity, but in the other it becomes confused, and the schist passes insensibly into it.

On Peel hill there occur beds of a similar description; also between Port-le-Murray and the small patch of limestone to its south, and in the floor of a quarry at Port Eshee, may be seen a bed of a similar quality, and two others which apparently spring from it, intersect the slate in the manner represented by fig. 3. One of these is bounded by soft clay on one side and indurated on the other, whilst the other has the hard clay on one side, but on the opposite is in contact with the slate itself. These last mentioned beds separate into concretions bounded by remarkably plain surfaces, differing from each other in form, but which do not strictly resemble any geometrical figure. Indeed all these beds appear to possess the same property to a certain degree.

DYKES IN THE KILLAS.

Dykes, which may be termed Elvan, occur in the island. One, two feet broad, is seen near the Dun, rising through the chlorite slate, a short distance below the junction of the two streams where, it was stated, the granite becomes concealed It is very variable in its composition, sometimes entirely quartz and chlorite.

On ascending Slieu-ny-Clough from Foxdale, along the bed of the river, there occurs a streak of granite, about one foot and a half wide, in the mica-slate which forms the west bank. No trace of it is seen on the east, but it agrees in character with a bump of granite to which it can be traced in the middle of the stream. It appears to affect the fracture of the greywackè bed above mentioned at Port Eshee. This consists of grains so exceedingly comminuted that we should scarcely be led to suspect its real nature from hand specimens.

LIMESTONE.

Dr. Berger has extended the limestone in his map too far to the north; the boundary lies in a direct line from Cass-ny-Hawin to Balla Salla. That which occurs at Port-le-Murray is also separated from the principal bed by an intervening patch of clay-slate.

In the immediate vicinity of the hill to the south of Athol bridge, the limestone encloses nodules of quartz, mentioned by Dr. Berger, p. 44. This may be traced thence, forming a broad band between Balla Hut and Balla Salla, and the appearances it presents lead me to reject his hypothesis, " that the existence of these concretions may be best accounted for by way of crystallization." I can conceive no possible objection to their being considered embedded pebbles from the shores of the sea, any more than the marine exuviæ, so prevalent throughout the limestone, to have been derived from the depths of it; but, if we examine the facts presented, I think but little doubt can be entertained of their being the worn fragments of older rocks.

In the first place, with the smaller nodules are mixed several pebbles, of large dimensions, whose structure is by no means crystalline, but granular; these again are associated with angular masses of quartz traversed by cotemporaneous veins of the same substance. The embedded materials when entering largely into the composition of any stratum appear pretty equally diffused, but when their number is comparatively small they occupy the lower portion, as if they had subsided from their greater specific gravity through matter in a liquid state.

A similar aggregation takes place in the lowest stratum of the small patch of limestone which sets on to the south of Port-le-Murray, and if I recollect rightly at Cass-ny-Hawin also.

Dr. Berger does not notice a change which sometimes takes place in the limestone, where its colour becomes reddish-brown and the texture crystalline. This appearance is slightly visible in Castletown bay, but occurs very plentifully to the south of Poolvash. Whether it forms a separate bed, in the latter case, or is merely a modification of the regularly stratified limestone is not so apparent. At one spot however, (marked *c, c*, in fig. 6, plate 35,) near the black marble quarries at Poolvash, two or three eminences occur of this nature rising through the regular strata, which are wrapped round and abut against them in a very perspicuous manner, the change being extremely sudden; the unstratified portion having a rugged appearance and being filled with fossils, the surrounding strata thin, slaty, and scarcely containing a trace of any.

For the characters of the annexed list of fossils I am indebted to Mr. Sowerby, who kindly undertook their examination.

1. Anomites punctatus.
   Martin. Tab. 37. fig. 6.
2. Producti. Min. Con. 59.
3. Productus Scoticus.
   Min. Con. 59. fig. 32.
4. Productus, finely striated.
5. Productus, another variety with intermediate striæ.
6. Cardium, new? See Min. Con. 82.
7. Spirifer. Min. Con. 265.
   Lin. Trans. Vol. 12. p. 516.
8. Spirifer, flatish.
9. Trilobite.
10. Nautilus complanatus.
    Min. Con. 261.
11. Ammonites Henslowi.
    Min. Con. 262.
12. Pentacrinites.
13. Madreporites. One of these very large and entirely new.
14. Tubipore.
15. Entrochi.
16. Caryophyllea. Several Species.

A dark shale, containing much mica, alternates two or three times with the limestone between Derbyhaven and Cass-ny-Hawin.

SANDSTONES AND BRECCIAS.

The sandstones pointed out by Dr. Berger, differ very considerably in character.

The sandstone at Peel, which alone is stratified, occurs both fine grained and under the form of a breccia; spangles of mica are dispersed through it, and it is sometimes mixed with much argillaceous matter. Strata of these appearances alternate. Towards the northern extremity of this formation, near Knocksharry, it terminates in an unstratified breccia containing limestone, agate, and ferriferous carbonate of lime. The limestone is of a dusky hue, very compact, and often an abundant ingredient. The agates are sometimes of a large size and very beautiful texture, but in general so full of flaws as to render them useless to the lapidary. Though they may be picked out from the sandstone, their union with this appears once to have been more intimate than at present, from a circumstance which happens with respect to the thin veins of crystallized carbonate of lime which traverse the bed in every direction: these invariably pass through every substance which they meet with in their course, even the smallest agates; from whence it would seem that the narrow fissures, into which this infiltration has taken place, were caused by the shrinking of the mass, during its consolidation, at a period when the texture of the different ingredients was more homogeneous than at present. If a crack of this nature were to take place now, it would most assuredly be turned from its regular course upon meeting with those agates, pass round them, detaching one side from the matrix, rather than traverse them in the manner described.

This bed is here seen penetrating and mixing with the clay-slate. Dr. Berger has committed a great mistake in supposing the whole isthmus of Langness to be composed of red sandstone. This is confined to the beach, and a small space inland, down the western side; the remainder of the isthmus being composed of clay-slate. The appearance also is totally different from that at Peel, and is that of an exceedingly rude conglomerate of quartz, slate, &c. in masses one or two feet in diameter, mixed with smaller fragments, and held together in a loose manner by a ferruginous cement. Wherever the removal of a portion of this bed, by the influence of the tide, had afforded an opportunity of examining the slate beneath, *this* appeared to have been rent and split in every direction into fragments, and these again cemented together, so that those belonging to any particular lamina may be traced for some extent, though thus confusedly disposed.

If I mean the same as that to which Dr. Berger alludes, when he says there is a sandstone overlying the limestone along the Castletown river, it is a breccia, which forms a hill immediately to the south of Athol bridge, consisting of angular fragments of quartz in a basis resembling that at Langness, but the ingredients are not so large or the texture so loose.

A breccia composed of angular fragments of limestone and quartz pebbles, loosely cemented by sand and calcareous matter, overlies the limestone at Caal Ferrane. This bed is not seen on the shore below high water mark, owing to the facility with which it becomes disintegrated by the tide.

The sand and gravel, which compose the high cliffs of diluvial detritus, towards the north of the island, are in many places agglutinated into sandstones and breccias by a calcareous cement which appears to pervade the whole of this tract to a great extent, so much so, that the roads, repaired with sand from this quarter, soon become hardened after a few showers. Fragments of shells and some nearly perfect specimens of recent cerithea, often occur sticking in the sandstone, and probably the whole of the calcareous matter here may be derived from this source.

AMYGDALOID.

From what Dr. Berger says of this formation, we might be led to suppose that the whole space between Scarlet and Poolvash was composed of a bed of amygdaloidal trap, which is not the case. The great mass of the formation is a trap-tuff of loose texture. Towards Scarlet it rises higher, becomes more compact, and acquires the character of an amygdaloid, the basis of which varies considerably in colour, assuming every intermediate shade between light grey and the deepest green. This portion appears to traverse the tuff like a broad irregular dyke, rising at intervals in abrupt masses, the highest of which, denominated the Stack, is insulated, except at low water. It is in several places split by numerous vertical fissures.

Fragments of the amygdaloid occur entangled in the tufaceous tract, as well as large nodules of the dark limestone, and even portions of its strata to a considerable extent, more particularly towards the south, where they have caused the appearance of an alternation; but I could observe no instance where the limestone might not be traced to an end, and be observed to feather out, with the evident character of an isolated mass. On examination we find that the embedded nodules, which occur so plentifully in the tuff, have been penetrated by it, and that it appears in little enclosed patches within the whole extent of their surface. Small cavities like air blebs also occur. These nodules do not appear to have lost their original colour or texture. Where the amygdaloid comes in contact with the limestone on the north, the latter for a few feet appears confused, is not stratified, contains many large cavities coated with bitter spar, and is intimately united with numerous small angular fragments of quartz, which are often exposed on the surface from the weathering of the rock.*

* I am aware that Dr. Mac Culloch has given a different view of this tract, since this paper was presented to the Society, but having again visited the spot I am inclined to think that he has taken his ideas from supposing the portion of limestone marked (*b*) to extend further than it really does. Perhaps the impossibility of visiting this rock, which can only be approached at very low tide, may have led to the error.

I observed five dykes in this neighbourhood, three to the north and two to the south of the tuff. The composition of all these is nearly the same, consisting of a tolerably close grained basalt, the northernmost containing a great quantity of olivine. Fig. 4, Plate 31, represents the horizontal section of this dyke formed on the beach. They all intersect the stratified limestone, and one also cuts through the tuff. They do not appear to alter the nature of the limestone, except in one or two places, where it has become white and charged with a quantity of argillaceous matter, but at the immediate contact it is filled with specks of basalt and crystals of hornblende, which are entirely embedded in the manner already mentioned with respect to those portions which have been entangled in the tuff.

Explanation of the Sketch, fig. 6, Pl. 35.

a, basaltic dykes.

b, confused mass of crystalline limestone, mixed with fragments of quartz.

c, crystalline limestone rising through the regular strata, containing many fossils.

d, the more compact portion of the trap, which is amygdaloidal and porphyritic.

A dyke, similar in character to these, occurs at Brada, running parallel to the course of the metallic vein along the coast to the north of the Head.

The strata towards the southern side of Langness are much intersected by dykes of greenstone.

A bed of greenstone, in which the hornblende is remarkably well crystallized, occurs to the south of Port-le-Murray. It may be seen on the beach, a few yards to the north of the small patch of limestone there, from whence it runs inland in a westerly direction forming a ridge eight or ten feet broad. It rests on the schist, and thin veins project from it downwards. There is a bed of trap resembling the more compact varieties of that at Scarlet, but not amygdaloidal, lying to the north of the Peel sandstone. It rises from the sea, and penetrates the schist in the form of dykes.

Hornblende slate is seen behind the Garth, near Kirk Marown, but whether in contact with the gneiss there, I was not able to determine, owing to an intervening spot of cultivated land.

SIMPLE MINERALS.

I have but little to add to Dr. Berger's list of simple minerals.

Fibrous Actinolite occurs in a decomposing state near the Dun, in two veins, each about six inches broad, traversing the decomposed portion of the granite and gneiss. It is accompanied by quartz, which it penetrates and frequently colours. It may be taken from the vein in fibrous bundles of three or four inches in length, but it is in general so much decomposed as to have assumed an earthy form. In this state it is mentioned by Dr. Berger, p. 54, under the name of Dun earth. On pressing the fibres between the fingers they crumble to a harsh powder capable of taking away the polish from glass, and consequently very unfit to be used in cleaning plate, a purpose to which it has been applied. I found a single specimen in which the fibres were flexible. Specific gravity$=3,03$.

I met with a worn lump of the arseniate of lead among the rubbish of the Laxey mine.

Veins of chlorite occur in different parts of the island traversing the clay-slate, and often colouring the quartz by which they are accompanied.

DILUVIAL DEPOSITES.

Curragh is a name given, in the Isle of Man, to any tract of peat bog. A large extent of land, called *the Curragh*, now under cultivation, and lying to the north of the mountainous tract, formerly consisted of this substance.

Much discussion has been excited in the island concerning the origin of its peat, some affirming that it is daily formed by the decay of certain aquatic plants, whilst others attribute its existence to the ruin of ancient forests. Those who favour the former hypothesis draw their conclusion partly from the phenomena observed upon cutting it. The method pursued is to take off the surface, which consists of vegetating matter, and having dug a trench to the depth of a few feet, to replace this on this bottom, when, in the course of time, this trench becomes filled up by a fresh body of peat, a circumstance attributed to the growth of the vegetable that is found on the surface. The opposite hypothesis accounts for the trench becoming replenished, by supposing the peat which forms its walls and base to yield to the pressure existing on all sides, and thus gradually filling up, in the manner that a ditch becomes obliterated when dug in soft mud.

It is not my intention to enter into an examination of the merits of either hypothesis, nor have my sparing observations on this head at all qualified me for such an office. Perhaps to the union of the two causes may be attributed the formation of the curraghs in the Isle of Man; but certain it is that some great catastrophe has here overwhelmed large tracts of forest land, and I am indebted to H. R. Oswald, Esq. of Douglass, for much of the information I received respecting the nature of these and the various trees which are found buried in them.

These curraghs exist both in the low lands and on the tops and sides of mountains, are very variable in their depth, and in most of them are found trees of a considerable magnitude, principally oak and fir; hazels also, with their nuts, are common, sometimes ash, walnut and black alder are met with, and in one instance holly has been found in a mountain curragh, tolerably green, but rotten and incapable of bearing exposure to the atmosphere. These trees never assume the appearance even of a partial conversion to charcoal, but the more they lose their original structure, the looser and softer they become. The oak is frequently tinged of a deep black colour, and is liable to split and crack when brought to the air, unless carefully seasoned by a gradual exposure; a large trunk will require two or three years training before it can be wrought with any degree of safety for purposes of ornament, but will then admit a polish no way inferior to ebony.

Towards the north the trees lie in general upon a bed of clay; some in the deeper curraghs are still erect; a few are torn up, but the generality are broken off about two or three feet from the roots, and are disposed parallel to each other, with their heads lying towards the north-east; any casual deviation from this position may be attributed to some obstacle, as the branches having changed the direction of their fall. The smaller boughs have either totally disappeared or lie crushed together in one mass.

This general disposition of the trees was long ago noticed by Sacheverill and Bishop Wilson, and is still fully believed to exist by the inhabitants, but I had no opportunity of examining this myself.

Between twenty and thirty years ago, during a violent storm, the sea laid bare a forest, about midway between high and low water-mark in Poolvash Bay, which had been covered with sand. It remained thus exposed for two or three days, and was seen by several people. The trees were of the pine tribe, and lay disposed in the same direction as those before mentioned, from south-west to north-east. Tradition reports a fine meadow to have been washed away hereabouts, by the encroachment of the sea at a very distant period.

In addition to these general remarks upon the curraghs, I shall mention some observations I made in the parish of Kirk Balaff, upon a substance which the farmers use there as a marl. It is confined to a spot within the compass of a few acres, and in it are found the remains of the gigantic Elk. Over this marl, to which I shall presently revert, is a bed of sand six feet thick and of a light colour; upon this lies a kind of peat, about a foot and a half thick, composed of rotten leaves, and small branches closely matted together, mixed with sprinkles of sand and containing a vast number of the exuviæ of beetles, bees and their nests, crushed together with seed-vessels, rotten, but having their external coating well preserved. This bed has not the wet and thoroughly decomposed aspect generally to be observed in peat; probably from the sand beneath absorbing the rain and moisture, and thus enabling it to remain in a state of unusual preservation. In general the hard wings are the only parts of the beetles which are preserved, and these are in appearance as fresh as on a living insect. Dr. Leach was enabled to identify a few with species at present existing in England. Upon this bed lies turf and vegetating matter to the depth of one or two feet.

The marl which lies beneath the sand is very different in appearance from the substances generally known by that name. It is white or greyish, and the fracture resembles that of a highly decomposed peat, though rather more earthy, and through it are dispersed in every direction the traces of small branches and roots, which partake of the character of the mass, and only become apparent where a light or dark line has traced their figure; in short a better idea cannot be conveyed of the appearance presented, than by conceiving a mass of peat converted to a loose, earthy, calcareous substance. When fresh dug it is moist, and possesses a peculiar odour, somewhat resembling that of hydrogen, which it loses upon exposure, and though it retains its texture, does not become indurated, but is easily reduced to powder between the fingers.

Upon connecting the several appearances here presented, it seems probable that an extensive forest was formerly overwhelmed by an incursion of the sea, bringing with it a large portion of sand. This sand would be sufficient to crush and keep down the main body, but some dry branches and leaves would rise to the surface and together with the insects form a scum there, which on the subsidence of the waters would remain resting on a bed of sand.

It would appear also that a herd of Elks must have here perished, so numerous are the remains which have been discovered within a small compass; but at the same time many of their bones have disappeared, if we may judge from the few which accompany the head and horns of each animal.*

Native phosphate of iron, in an earthy and pulverulent state, is sometimes found in contact with the bones; a similar substance I have observed coating the external surface of the empty shells of the Mytilus cygneus of Linnæus, which lie scattered over the fens of Cambridgeshire, after the crows have extracted the animals.

---

* Whilst I was on the island, in August, 1819, two heads with the branches and a vast number of bones were dug up in the finest preservation. An ingenious blacksmith in the neighbourhood, taking the skeleton of a horse for his model, has contrived to put together these bones with great accuracy, and form a skeleton in which the only parts wanting are the half of one hoof and the end bones of the tail. It has a most magnificent appearance, stands six feet six inches to the top of the back, and thirteen feet to the tip of the horns.

From the position in which the trees are said to lie in the curraghs, it would appear probable that some overflowing of the sea, in a direction from south-west to north-east, took place at a very distant period, but the few observations I made on certain banks of diluvium in various parts of the island do not countenance this idea. Some of these occur along the road from Peel to Douglass, and thence towards Laxey. They are generally composed of gravel and sand, and contain rolled blocks of various descriptions and sizes disposed in horizontal or wavy lines. Those about Douglass contain, principally, fragments of quartz and clay-slate, but further north we meet with a few blocks of granite, at Laxey, with very many, and between Laxey and Dun, the stone fences, which have been collected from the fields in the neighbourhood, are composed of little else. This granite is of the same appearance with that at Dun, and though dispersed so abundantly to the south of that place, I did not observe any towards the north. In an undisturbed bank of diluvial detritus at Clovenstone, I found a large block of quartz traversed by crystals of actinolite similar to those met with at Dun, and another block in a similar situation about two miles to the west of Douglass.

Blocks of granite, similar to that of Slieu-ny-Clough lie dispersed for two miles over the low plains between St. Mark's chapel and South Barrule, nor have they the least appearance of resting on their birth-place, so that no cause appears so likely to have placed them there as that which scattered those of Dun. This disposition of the granite boulders appears to indicate a current having set over the country, (at least on the eastern side of the central chain) in a direction from north-east to south-west, or contrary to that pointed out by the alleged disposition of the bog trees.

Beyond the northern extremity of the mountainous group the

whole country consists of diluvial matter, raised in many places to a considerable height above the level of the sea, excepting to the north of the Balla Chirrym hills, from whence there extends, to the point of Aire over two miles, a low sandy plain covered with rolled pebbles.

Dr. Berger has remarked, p. 34, that it is by some asserted, that the sea loses ground here at the rate of two yards per annum, but from enquiries made on this point, I am inclined to think this circumstance overrated. Though all who have lived long in the neighbourhood allow that there has been a very sensible increase of land within their memory, the mean rate at the utmost would not appear to exceed half that mentioned.

The cause of this increase may possibly be ascribed to the tides which act upon the island. These enter the Irish channel at two quarters, that from the south running up the island meets the northern between Kirk Maughold and the point of Aire, where their united efforts form the flood, and their strength being exhausted by opposing forces, we may imagine that they would leave deposited whatever each brought with it, causing an extent of shallow water for a considerable distance. It will readily be seen that the body of water raised over this deposite would not, on subsiding, produce the same effect in removing as a current in accumulating it; thus permanent additions would daily take place, and the portion contiguous to high water mark become gradually covered with soil and vegetation till it is at length removed without the influence of the tide. The numerous blocks of various descriptions and magnitude found along the north coast, are probably derived from some distant spot, for although several of these have doubtless fallen from the cliffs, yet their vast number precludes the idea of all having come from thence, more especially on the north-east quarter where the sea does not wash their base but in few places, being prevented from reaching it by a slight accumulation of sand and pebbles. But the blocks which are scattered through the high cliffs of sand and gravel were themselves most probably derived from the very same source as their neighbours on the beach.

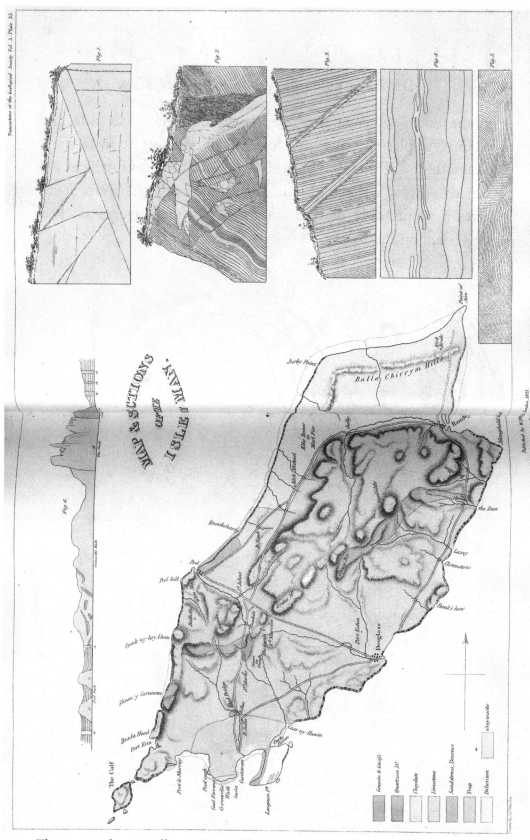

The material originally positioned here is too large for reproduction in this reissue. A PDF can be downloaded from the web address given on page iv of this book, by clicking on 'Resources Available'.

TRANSACTIONS

OF THE

CAMBRIDGE

PHILOSOPHICAL SOCIETY.

———

Vol. I. Part I.

———

CAMBRIDGE:

PRINTED AT THE UNIVERSITY PRESS,

AND SOLD BY

DEIGHTON & SONS, AND NICHOLSON & SON, CAMBRIDGE:
AND T. CADELL, STRAND, LONDON.

———

M.DCCC.XXI.

XXVI. *Geological Description of Anglesea.*

By J. S. HENSLOW, M.A.; F.L.S.; M.G.S.

ST. JOHN'S COLLEGE,

SECRETARY TO THE CAMBRIDGE PHILOSOPHICAL SOCIETY.

[Read *Nov.* 26, 1821.]

To accompany the present Memoir, I have formed a collection of the rocks of Anglesea, which has been placed in the Woodwardian Museum. This collection is numbered throughout, and the number corresponding to any particular specimen is noted between brackets, whenever any allusion is made either to its locality or to the nature of its composition.

I have to acknowledge my obligations to L. P. Underwood Esq., whose previous visits to Anglesea had enabled him to collect many interesting facts connected with its Geology, and to whom I am indebted for the locality of several trap-dykes, which might otherwise have escaped my observation.

I believe that no good map of Anglesea has yet appeared. The map which accompanies this paper is compiled from two maps of North Wales, one by Furnival, published in 1814, the other by Evans, in 1797. The first of these furnishes, with con-desirable correctness, the relative positions of the towns and general outline of the country, but does not pretend to trace the indentations of the coast. Evans has enabled me to give some of the latter, where they affect the geological details; but neither in this respect, nor in the configuration of the surface, could I procure any accurate information. What is here offered must be considered as a very rough approximation.

As the map is rather complicated, it has been thought advisable to adopt an artificial arrangement of the different districts in each formation. By this means a reference can more readily be made to any particular place, without the labour of searching through the several detached portions marked by the same colour. A table explaining this arrangement is placed with the description of the plates; and the references are made on the margin whenever they seem to be required. No other places are noted in the map than those alluded to in the paper.

In the accompanying sections, continuous lines are meant to represent the portions actually exhibited. The dotted lines are either such portions as were not visited, or were too much obscured by the concealed nature of the ground. Where no actual section exists, the junctions are marked by the dotted line, even where the boundary between two formations is sufficiently evident at the surface, and the order of collocation has been either ascertained along the coast, or cannot be doubtful from the characters of the contiguous rocks. As the sections are parallel to each other, a reference may readily be made to corresponding portions, and the spot seen where the order was clearly established.

Anglesea possesses no very striking features. Holyhead mountain, which forms the greatest elevation, reaches only to 709 feet above the sea. With this and two or three other exceptions, the ground is low and undulating, although the surface possesses by no means an uniform character. A further description may perhaps, with greater propriety, be referred to the details of each particular formation.

The plan of this paper will be, first to describe the stratified and then the unstratified rocks.

The greater portion of the stratified rocks has suffered considerable disturbance, and they frequently occur under characters very different from what they assume in their undisturbed state. Several of the details therefore, which would otherwise be included in this division, are deferred to the description of the unstratified masses; when it is to their intrusion that such phenomena can be referred.

Quartz Rock.

{Nos. 1. to 11.}

The term, Micaceous schist, would perhaps by some Geologists, be made to include the whole series of the oldest stratified rocks. These, however, vary considerably in composition, but do not allow of separation into distinct formations, and would scarcely admit of geographical distribution, even in a map of the largest dimensions and most accurate construction.

In one instance an exception may be made in favour of a variety, marked on the map as a quartz rock, which possesses certain peculiarities of structure, though it is not very remote in composition from other varieties included under the general denomination. It occupies two distinct localities, the one in the Q.1. northern division of Holyhead Island, lying to the West of a line drawn from Port-Dafreth, to a point on the shore about midway between Holyhead and the mountain. The other, in the Southern Q.2. division of that Island, occurs in the neighbourhood of Rhoscolyn, extending along the coast, and bounded by a line drawn from Borth-Wen to Rhoscolyn church, and thence about one mile further to the N.W. In each case this rock rises to a greater elevation than the surrounding country, and in the first-mentioned situation forms the highest point of all Anglesea.

On the summit of Holyhead mountain, and on the highest point near Rhoscolyn, the term given to this rock is strictly applicable, it being composed of little else than highly crystalline and distinctly granular quartz (1.) firmly cemented. There occur in it a very few minute white specks of earthy felspar. It is much intersected by cotemporaneous veins, and occasionally tinged red (3.). In other places it intermixes with a little mica (4. 5.) The quartz is often finely granular (6.), and associated with larger and distinct crystals of silvery white mica (7. 8.), and apparently also with a little chlorite, the specimens assuming a greenish tinge. Such specimens strongly resemble greywacké, but their crystalline nature is still very distinct.

With regard to the structure of this rock, nothing can be more deceptive than the appearance which it assumes in places Q. 1. where no extensive section exists. On crossing Holyhead mountain, we seem to be walking over the edges of parallel strata, which dip at a very high angle towards the N. of W., the whole surface consisting of broken rugged lines, running from S. of W. to N. of E. Opposed to the small island called the South-stack, is a perpendicular cliff of two or three hundred feet in height, exhibiting the structure of the mountain in a perspicuous manner. Every trace of the former apparent disposition of the strata is lost, and the whole is seen to consist of broad strata, contorted in a most extraordinary manner, often vertical in position, then returning with a sudden curvature, and forming repeated arches. Strings of white quartz, which occur in and between them, partake equally of these contortions, and also of others more complicated and independent of the general position of the surrounding mass. The effect is rendered still more striking by each stratum assuming a peculiar tint; the colours varying through obscure shades of green, brown, and yellow.

They also vary in texture, which causes the more compact portions to project in relief, and these in weathering exhibit convolutions in which the softer strata do not partake. They are sometimes divided by fissures, generally placed nearly at right angles to the curves, producing the effect of an artificial stone arch. Plate XV. represents about fifty feet perpendicular height of this section. Wherever a convenient opportunity occurs of examining the cliffs along the remainder of this district, the same appearance is repeated.

Near Rhoscolyn the strata are very distinct, and among Q. 2. them is one of a brick red, contrasted with others of a deep yellow colour. The same deceptive appearance of stratification running from the S. of W. to the N. of E., is seen in this district as well as in the former.

The real structure of this rock, then, consists of a succession of contorted strata rudely conformable to each other. That these were originally deposited in their present position seems impossible, and the whole bears a striking resemblance to the flexures that might be formed in a pasty unconsolidated mass, by the application of a disturbing force.

The deceptive appearance resembling stratification, arises from the parallelism preserved between the scales of mica dispersed through the mass, which causes an imperfect kind of cleavage throughout the whole district. In some cases this may readily be exhibited, even in hand specimens (6.), where the mica has a brown ferruginous aspect, coating over the whole surface produced by cleavage, with a plate too thin to be detached. These laminæ occur about one eighth of an inch asunder, and on fracture exhibit an uneven undulating surface. A few other small scales of white mica are irregularly dispersed through the specimen. In some cases (4.) the laminar tendency is distinct, with the intervention of a very small portion of mica.

and on a fracture perpendicular to this direction, its existence is marked by faint lines. In other instances the cleavages, sufficiently apparent on the large scale, would not be noticed in a small fragment. Here (7.) the mica which gives rise to them is dispersed at intervals over their surface, and is scarcely to be seen upon a transverse fracture.

Other cleavages exist in the mountain, at much greater intervals than the former, which more nearly resemble natural fissures. These present smoother surfaces, and are probably the cause of large vertical fissures which occur towards the summit of the mountain, separating the rock into rude rhomboidal and cubical masses, where quartz is almost the sole ingredient, and both the fissile texture and divisions of the strata are scarcely to be detected, though, in convenient situations, each may occasionally be traced.

The exposed faces of these masses (2.) present an even polished surface, the effect of weathering; an action which apparently soon ceases, the fragments retaining their angles as sharp as when first fallen.

Plate XVI. Fig. 1. represents the positions of four cleavages obtained from a projecting mass of curved strata (6.).

A. The curved surface of the stratum.

$$
\left.
\begin{array}{rcl}
1 - 2 &=& 52^{\circ} \\
1 - 3 &=& 90 \\
1 - 4 &=& 120 \\
2 - 3 &=& 90 \\
3 - 4 &=& 136
\end{array}
\right\}
$$
Rough calculations of the angles at which the several planes are inclined to each other.

Plane 3. is the cleavage which produces the apparent stratification of the whole mountain.

It should seem that these phenomena arise from some effort of crystallization subsequent to the original deposition of the materials, and subsequent also to the present contorted position of the strata. These facts may be illustrated by the appearance presented upon the transverse fracture of a calcareous stalactite, where the original structure arising from successive depositions, is exhibited by concentric circles, whilst a rhomboidal fracture marks the effects of a posterior crystallization. The strict resemblance which some of the strata bear to those of sandstone, points out a mechanical deposition as the most likely mode of formation: their present structure may suggest an idea that crystalline force assisted by moisture and pressure is an agent of sufficient power to have produced the similar, but still more perfect texture of the oldest stratified rocks, without the necessity of imagining any previous solution of the ingredients which compose them.

Chlorite Schist.

{Nos. 12 to 187.}

Under this denomination are included several varieties of schist, of which quartz and chlorite form the principal ingredients. Mica slate and clay slate also occur in the same formation, but each of these passes into chlorite schist by insensible gradations, and I could no where trace a boundary between them, marked either by a rapid change in the mineral character, or by some distinct geological feature. The varieties of clay slate included in this formation appear to consist of nearly the same ingredients as the chlorite schist, and to differ from them in nothing but want of crystalline structure. They are of various shades of green and red, and of a close texture.

The variety which immediately succeeds the quartz rock is crystalline chlorite schist. There are four sections on the coast

C. 1. exhibiting their union. At the spot where this takes place between Holyhead and the mountain, there is a space of eight or ten feet in width, occupied by a rock intermediate in character between the two. At the spot on the beach where the change has become decisive, there occurs a breccia (9. 10.) of quartz rock, and angular fragments of talcose slate; from this a vein issues, which may be traced on the shore for a few feet as far as the cliff, up which it is seen to rise, and become forked near the top (Pl. XVI. Fig. 2.). The vein consists of finely granular materials (11.), with occasional patches of the breccia from which it proceeds. Possibly, the origin of this vein may be ascribed to a fissure in the chlorite schist having been filled from the subjacent rock, previous to its consolidation, by the force of the superincumbent pressure.

C. 1.   At Port Dafreth, I could not ascertain whether any sudden change takes place; the strata of the quartz rock gradually be-
Q. 2. come thinner, and appear to pass insensibly to chlorite schist the West of Rhoscolyn their boundary is more marked, the quartz rock rising nearly vertically, and the chlorite schist resting unconformably against it. A small bay is there formed by the removal of the chlorite schist, so that its Eastern side is composed of quartz rock alone, where a broad stratum presents an undulating and nearly vertical cliff. At certain points of projection, a portion of this is removed, which exposes the stratum next below; a similar exposition of the next takes place, and so on. The nature of this may perhaps be better understood by referring to Pl. XVI. Fig. 3.

The last spot where this junction takes place lies to the South-east of Rhoscolyn, and West of Borth-Wen, where the confusion is greater than at either of the three other places. The quartzose strata intermix with the chloritic, and are seen on the beach like veins twisting among them. In some instances they even appear to occur in an inclined position above them.

From what has been said of the nature of these junctions, it may be a question, whether the quartz rock and chlorite schist are not members of the same formation. For, whilst the junctions at each of the Western boundaries of the former bespeak a nonconformity, those on the Eastern sides present a certain degree of intermixture. Supposing the quartz rock to have been once horizontal, and the chlorite schist reposing upon it, perhaps their present appearance may be accounted for upon the principle of an upheaving force acting obliquely from the East.

Around Holyhead the chlorite schist is greenish grey, and C. 1. composed of nearly equal proportions of quartz and chlorite, both highly crystalline, and finely granular. Sometimes it possesses but an imperfectly fissile texture (12.), though in general this is sufficiently apparent (14.), and the quartz and chlorite predominate in alternate layers (15.). The scales of chlorite often form one continuous shining plate on the plane of cleavage (13.). There are still finer grained varieties (16.), where the intermixture of the quartz and chlorite is very uniform. These form an intermediate passage between the crystalline chlorite schist, and bright green silky clay slate. Quartz sometimes predominates considerably (17.).

Contortions of the most complicated nature are exhibited in many portions of this series. A large block will often present laminæ waving in regular vandykes, or intermixing in a most confused manner (18.).

The finer grained varieties appear to predominate through the remainder of the chlorite schist situate towards the Western side of the Island. The general character of the same rock on C. 5. the Eastern side is slightly different, though the composition is similar. Both the grains of quartz and the scales of chlorite are larger, and the colour of a darker green. Lenticular plates of

pure crystalline quartz, lie in the direction of the layers of chlorite, which pass regularly round them (19.).   Strings of crystallized quartz and scales of chlorite intermix in irregular layers
(21.), which appear to arise from the complicated contortions of
thin laminæ. Sometimes the rock is nearly homogeneous, the
chlorite being dispersed through the quartz (23.).

C. 5. Some hard varieties occur in the neighbourhood of a rock
upon which there is erected a pillar to the Marquis of Anglesea,
near Plas-Newydd. The basis is fine grained quartz, and dark
grey chlorite (24.) closely united, and through the mass are
disseminated small crystalline patches of light yellow epidote,
and others of reddish mica.

The epidote, in some cases, forms a considerable ingredient
(25.), and the scales of chlorite are replaced by dark green
spiculæ, probably hornblende.

C. 5. Mica slate occurs in the Eastern chloritic district, which
lies to the S.E. of a line drawn from Pentraeth to Newborough.
This is, however, always contaminated with some portion of
chlorite, which may be detected by the earthy smell, even of
the purest specimens.

The hill West of Penmynydd, about the centre of this district, on the main road from Bangor to Holyhead, and the
Llydiart mountain, at its N.W. termination, afford the most
crystalline and genuine examples of this rock (26—29.). The
quartz does not always present a distinctly granular appearance;
but rather constitutes a nearly uniform mass, through which the
scales of mica or chlorite are dispersed. It is not always easy
to determine whether mica or chlorite forms the second ingredient. Sometimes both are present, and sometimes the quartz
is tinged green by an intimate mixture with the chlorite, and
scales of white, or light green, mica are superadded.

On the shore, between the Menai bridge and Plas-Newydd,

mica enters in a conspicuous manner as an ingredient of the schist, and forms large thin plates which are irregularly inter-mixed with an impure chlorite schist (30—32.).

Near Cadnant the scales appear to be intermediate between mica and chlorite, and coat over the whole surface of cleavage (33.).

Other varieties afford an impure mixture intermediate between clay slate and mica slate, where the scales of mica are sufficiently distinct, but the basis no longer retains a crystalline character (35—37.).

North-east of Bodwrog, on the confines of the granitic c. 3. district, is a variety composed of crystalline white quartz in layers, coated with a talcose variety of mica of a light straw colour (38, 39.). Mica slate occurs to the South of this spot, along the Eastern boundary of the granite, where it does not present a granular aggregate of quartz and mica, but forms a highly crystalline mass of quartz, to which a laminar tendency is given by thin layers of mica, sufficiently distinct on the surfaces of the laminæ, and but faintly marked by lines on the fracture perpendicular to them. Where the rock is weathered smooth, the quartz glistens in the same manner as a polished surface of crystalline marble (40.).

A similar variety is found on the East of Tre-Sgawen (41.), c. 3. situate towards the North-eastern termination of this district.

The passage from the crystalline varieties of chlorite schist to the more earthy kinds (43—49.), and finally to clay slate (50—64.), is very gradual. The yellow epidote, before-mentioned, also assumes a compact appearance, and runs in irregular strings among the schist (43, 44.).

Some specimens of the clay slate present a silky lustre (50), and fibrous appearance. Patches of deep red occur inter-mixed with the green.

C. 2.   These varieties of clay slate predominate to the North of a line drawn from Llaneilian to Llanfechell. On the coast, from
C. 1. Llanrhyddlyd to the nearest point to Holyhead Island, they intermix with the crystalline varieties of chlorite schist, in the most confused manner. A boundary may once have existed between them, for the transition from one to the other is frequently abrupt, and resembles a series of patches of clay slate scattered over a ground of chlorite schist, sometimes presenting a distinctly laminar tendency, at others not a trace.

The clay slate often assumes a hard jaspideous aspect (59-61.). Some earthy varieties, without a trace of fissile texture, appear to consist of an irregular mixture of chlorite and epidote, with patches of quartz, and carbonate of lime (65—69.), intimately
C. 5. united. These prevail on the coast from Beaumaris to Cadnant, in the neighbourhood of Llangaffo, and also between Trefdraeth
C. 3. and Aberfraw.

Rugged, projecting masses of schist, with laminæ generally much contorted, are scattered throughout the tract on which
C. 1. Holyhead is situate. The average bearing of the laminæ is decidedly towards the N.E. and S.W., and their dip in general to the N.W.

C. 2.   In the Northern district round Llanfechell, the appearance is similar, but the laminæ are not so much contorted, and they dip more to the North. Towards Amlwch they nearly regain their former position, and between this place and Llaneilian, the contortions are very complicated.

Where the rock had been vertically and smoothly cut, in a recent excavation opposite to the pier at Holyhead, there was a decided appearance of broad strata, undulating in a manner similar to those of the quartz rock, and also a laminar structure parallel to the seams which marked the stratification.

A similar circumstance occurs in the high ground to the C. 2. North of Llanbabo; the surface is modified by the undulating nature of the strata, which rest upon each other conformably, and are from two to three feet thick. The schist is very flinty, and possesses but little appearance of fissile texture.

In the most Easterly tract of this formation, the denuda- C. 5. tions inland often appear in small rounded eminences, with smooth surfaces dipping gradually on one side, and presenting a vertical face on the other. With a little attention these are distinctly seen to be stratified in the manner represented (Pl. XVI. Fig. 4.), which is intended for a section of the Pillar rock near Plas-Newydd, and an eminence immediately on its N.E., when viewed from the East. This character prevails on each side of a line drawn from Llandonna to Newborough: but on the coast, from Beaumaris to Cadnant, the dip is South-easterly.

Where a separation into laminæ does not exist, the scales of mica or chlorite still preserve a degree of parallelism in the crystalline varieties, which, combined with the curvature of a stratum, produces irregular but parallel lines upon its exposed surface, whose general bearing is still towards the N.E. and S.W.

From these circumstances we may perhaps conclude, that wherever a laminar tendency is found in this formation, it was originally parallel to the planes of stratification. And here there appears to exist a marked difference between this and the quartz rock, in which it should seem, that the laminar tendency has arisen from an arrangement of the particles posterior to the present contorted position of the strata.

In endeavouring to account for any appearance exhibited by these rocks, it is necessary to take into consideration the more nearly homogeneous nature of the chlorite schist, when compared with the very variable strata of the quartz rock.

In some places the chlorite schist is associated with rocks composed of heterogeneous materials confusedly aggregated (70—94.). The schistose character is more or less destroyed, and the argillaceous basis intermixes with crystalline limestone, dolomite, C.3. serpentine, and jasper. The largest tract of this description lies about one mile to the West of a line drawn from Llangefni to Trefdraeth. The ground is completely broken up by rugged projecting rocks. Some of these are slaty, but in general they present a hard jaspideous aspect with contorted stripes, which mark the existence of former laminæ. These intermix with homogeneous red jasper streaked and spotted with purple.

The width of this tract may be about one mile, and it is succeeded on the West by contorted chlorite schist, and this by the mica slate already described, which, though confused, decidedly dips from the granite.

C.4.   A similar character prevails in the small detached strip of chlorite schist, which forms a ridge from Caint to Red-wharf bay, passing between Pentraeth and the Llydiart mountain. From Caint, as far North as Llanffinnan, this ridge is composed of green and red glossy talcose clay slate; but immediately North of Llanffinnan, it becomes disturbed, passes to a compact red jasper (58, 59.), and from hence to Red-wharf bay presents a series of broken elevations, composed of fragments of schist cemented by crystalline magnesian limestone, patches of which, as well as of compact limestone and jasper occur through the remainder of this district, intermixed with schistose materials. At its termination in Red-wharf bay, it forms a low but perfectly vertical cliff, facing the N.W., intermediate between red jasper and clay slate (60.), and possessing a fissile texture. Both the common and magnesian limestone present different tints of grey, yellow, and flesh red (88—93.).

In contact, and on the steep side of one of these projecting masses of limestone, is found a calcarious tuffa (94) enclosing fragments of slate, and recent snail shells. I mention this circumstance as it may perhaps tend to shew, that some particles of this limestone have been in a state adapted to solution at no very distant period, although its present position should seem to indicate, that this action has ceased.

Fibres, resembling a coarse asbestos, penetrate the solid jasper (86.), and sometimes appear as small veins, (the fibres perpendicular to the sides) traversing a light porous mass into which the jasper passes.

At the Southern point of the promontory at Llanddwyn, c. 5. there is another partial formation of a similar nature. At one spot are numerous kernels, about the size of peas, dispersed through the schist (80, 81.). These appear to consist of a light green serpentine, in which lime predominates.

Half way between Beaumaris and Garth-ferry, in the new road, there is a rude projecting mass of rock, composed of red crystalline limestone, and jasper (83.), embedded in, and intermixed with decomposing argillaceous materials. When passing close to this, it appears to form a high projecting point of the cliff, but viewed from the river, it is seen in reality to be situate in the bottom of a gap formed by the schist rising abruptly on either side.

The clay slate, on the S.W. slope, and near the summit of Bodafon mountain, situate at the Northern termination of the c. 5. middle district, passes to a compact mass between hornstone and jasper (95—100.). It is irregularly streaked with different shades of green, dull red, and grey. It fuses to a transparent frothy white glass, and probably contains a great proportion of felspar. Indeed the red stripes actually pass from a compact nature to crystalline veins of felspar, which are occasionally associated with stripes of white quartz. Specks of sulphuret of copper are dispersed through the mass.

Contorted patches, and strings of crystallised quartz and red felspar (121—123.), occur in several parts of the chlorite series, both among the crystalline and earthy varieties. In the M. 1. new road to Holyhead, S.E. of Llanfihangel-East*, I procured masses of crystallised felspar four inches cubed (123.). The colour varies from deep to light red; the structure is curved-laminar passing to compact.

C. 5. On Red-hill, to the South of Beaumaris, is a bed of crystalline quartz (118, 119.), which is quarried for the Staffordshire ware. Other beds of a similar nature are met with in various parts of this schist.

On the shore at Cadnant are broad veins of quartz, slightly contaminated with chlorite (120.). These veins pursue a direct course, and resemble trap dykes in external character.

G. 1. A broken flinty ridge runs from the N.E. side of the Paris mountain to the S.W. of Llaneilian mountain. A fissile texture is sometimes visible, and the rock passes to a schist (111.). Its fracture and aspect vary from flinty to cherty, and its colours are different shades of light green (113, 114.), grey (115.), and red (116.). Sometimes there are small crystalline specks of quartz and felspar dispersed through the mass (117.), which give it a porphyritic aspect. It is semi-translucent on the edges and fuses to a white frothy enamel. This is, perhaps, more nearly allied to hornstone than that from Bodafon mountain.

The transition to a compact flinty or cherty mass, is found in several other portions of this district (108—110.).

---

* I have added "East" to the name of this place, to distinguish it from another Llanfihangel situate on the Western side of Anglesea, on the confines of the chlorite schist (c. 1.) to the S.E.

Limestone Beds.

{Nos. 124—132.}

Limestone, in the form of veins and small patches, has already been noticed; it also exists in distinct irregular beds in several places, which are marked in the map by an L. In the cliff, East of the island on which Llangwyfan church is situate, C. 3. there is a bed of compact white marble mottled with black (124, 125.).

Other beds of the same nature occur on the promontory South of Aberfraw.

Also at Gwalchmai, immediately to the S.W. of the lake. As the limestone passes into the schist, it assumes a fissile character, and scales of chlorite are dispersed over the natural fractures (127.).

A compact dark brown and grey limestone (131, 132.), not unlike some of the more crystalline varieties of mountain lime, has been quarried about Llanfacthlu to a considerable extent. C. 1. There is an impure shaly substance associated with it (132.), somewhat resembling the shales of the coal measures.

Small caverns occur in this spot, the surface of which are rugged, and contain hollow cavities resembling the exposed portions of a limestone district on the sea shore. No stalactites are to be found in them.

Very considerable beds of a similar limestone extend from C. 2. Glan-y-Don to Cemmes.

In none of these beds was I able to find any trace of organic remains.

Serpentine.

{Nos. 133 to 187.}

Two districts are laid down in the map, in which the principal masses of serpentine are found. These occur in beds subordinate to the chlorite schist, and do not form one continuous line of rock.

S. 1.

In the Southern district, they form a range of detached and nearly tabular masses, which extend from the N.W. of Rhoscolyn church to Llanfihangel, rising through swampy ground, and accompanied by projecting patches of schist which dip in various directions Pl. XVI. Fig. 5. The compact serpentine passes into slaty; and sometimes a tabular mass exhibits this double structure, when viewed at a short distance, Pl. XVI. Fig. 6.

S. 2.

The serpentine near Llanfechell is not sufficiently exposed to enable us to trace its connexion with the schist. The patches, in which it is found, have been quarried, and appear to be nearly envelloped by a hard compact variety of chlorite-schist.

The purest specimens are dark green with a semi-translucent greasy lustre (133—135.), but the general appearance is that of a compound rock, in which serpentine and dolomite form an irregular mixture (141—151). Patches of light yellow also occur (145.). A considerable portion of that which is quarried at Llanfechell, consists of very compact dolomite, tinged green (146.) or red (148.); sometimes striped (150.); in which patches of serpentine are embedded. It is here associated with common compact limestone (152.). The red tinge also pervades some of the more slaty varieties (153—155.).

S. 1.

Near Rhoscolyn the serpentine is associated with a heavy, compact and granular, black limestone, which does not resemble dolomite, although it will not effervesce in cold acids (161—163.).

Patches and veins of beautifully saccharine and white dolomite are dispersed through each district (159.). This occasionally exhibits a tendency to a fibrous structure (160.), which may sometimes be traced partially through several specimens of the serpentine.

Asbestos forms a thin coat over the natural fissures of the serpentine in the form of mountain leather. It also occurs in thin veins, of a light green colour; the fibres set perpendicularly to the sides of the vein (156.), which sometimes seems to be contemporaneous (136.). Some specimens appear to consist of broken fragments of this substance cemented in a paste of serpentine, in which the direction of the fibrous structure being inclined at different angles to the surface, a polished specimen (157.) has a beautiful appearance, different fragments reflecting the light at different angles of inclination.

Pyrites occurs dispersed through the serpentine (136. 140.).

Small crystals of jet black pyroxene form also a common ingredient (137—139.); but they are so intimately associated with the mass that they can not readily be detected, except upon a weathered surface, over which they are scattered in projecting points.

The schistose portion of the district in which the Rhoscolyn serpentine is situate, varies considerably in composition. On approaching the serpentine, asbestos enters largely as an ingredient. This is intermixed with slaty and chloritic serpentine irregularly laminated, with carbonate of lime (170.). Other varieties approximate to chloritic slate (164—169.). A structure half fibrous half slaty is a common character (171—177.).

Radiating crystals of dark green actynolite are dispersed through a more compact variety, the fibres generally lying on the surface of cleavage (180.).

In the confused schist along the shore South of Lanfacthlu, [c. 1] are several appearances which approach the character of the Rhoscolyn district, though no considerable mass of serpentine is seen. This schist is sometimes a mixture of serpentine and chlorite, and in it are beds and veins of compact limestone (186.) and earthy chlorite (187.).

Grēywacké.

{Nos. 188. to 264.}

Under this denomination is included greywacké slate, and also a fine grained dark grey or black clay slate, which cannot be distinguished in composition from the green clay slate of the last series. It exists, however, above and also intermixed with the greywacké, in a manner which decidedly places it in the same formation. Whether this class of rocks originally succeeded the former in an uninterrupted order, or whether they were separated by a marked geological epoch, cannot be fully ascertained in Anglesea. Their characters are sufficiently distinct to enable us to trace their boundary on the map.

In a few instances the black clay slate assumes a glossy crystalline appearance, approaching the character of a primitive clay slate (188—192.), which passes insensibly (193—196.) to the earthy varieties (197—207). It is often thin slaty, but the plates are not sufficiently regular to admit of their being wrought. The more. common character is that of a shattery schist, breaking into small irregular fragments. In several places this schist is intermixed with a greywacké conglomerate (210, 211.) consisting of angular fragments of slate, embedded in a fine black argillaceous basis, or it is composed of quartzose fragments with the addition of argillaceous matter (212—216.). The black clay slate intermixes also with a grey sandstone (221—229.) which cannot be separated from some of the sandstones of the old red sandstone.

G.1.   In the bed of the river at Dulas, the greywacké approaches a sandstone, and contains small embedded fragments of schist (209.). It cleaves into irregular laminæ about one inch in thickness, intermixed with others of a more shaly character. Between the laminæ are hard nodules (258, 259.), of a concretionary nature, composed of the same materials, which decompose in concentric crusts. The finer slaty laminæ pass round them.

On the Western side of Dulas harbour, the greywacké is intermixed with shaly black clay slate, which forms the greater part of the rocks to the S.W. Upon approaching its termination towards the North, where the conglomerate sets on, and where it is intersected by several trap dykes, the harder laminæ (226.) increase in number until they form the body of the rock, with a very slight portion of shaly matter interposed. Concretionary nodules are also found here which possess a peculiar structure (260, 261.). In shape they approach a spheroid, slightly flattened on one side. Upon examining the more convex surface, the nodules appear to consist of cylinders of different sizes pressed together, so that an imperfectly columnar structure is the result; the termination of the cylinders on the surface forming rounded projections. Upon fracture these cylinders are found to be composed of a succession of cones, each about one tenth of an inch thick, placed one within the other, with their bases towards the convex side of the nodule. The surface of each cone is irregularly wrinkled longitudinally, and marked transversely with faint striæ. One cone runs into another, and the whole is so blended together that it is impossible to detach a perfect cone from the rest. There exists a slight tendency to natural cleavage, inclined to the shorter axis of the nodules at an angle of about 45°, which is also about equal to the inclination of the conical surfaces to the same axis. The major axis of some of the larger nodules is two feet and a half, and the minor one foot and a half; and the conical structure extends to the depth of three or four inches. The direction of the longer axis is placed parallel to the schistose laminæ, which pass round the nodules. There is one hard lamina, fifteen inches thick, nearly vertical in position, which winds among the schist in a most irregular manner, closely resembling a basaltic dyke in external character.. It may be traced for some distance along the beach, and also up the face of the cliff. On one side it is completely studded with these concretions, but in this instance their form is modified, the side next to the lamina being flat, and the conical structure extending through the whole of each. They are generally separated from the lamina by a thin seam of clay, but are sometimes firmly united to it. The concretions are confined to one side of a lamina. These laminæ are frequently striated black and grey with all the regularity of a fine sandstone (226.). The broad one above-mentioned is uniform in character, and consists of finely granular and highly crystalline quartz and felspar, partially blackened by argillaceous matter (222, 223.).

G. 1.　　　In the denuded patches of this series round Llanerchymedd, the greywacké character prevails, the base being a black clay slate, which encloses fragments of quartz and slate. More to the East and N.E. it generally consists of shattery fine grained black clay slate, which is also found throughout the strip extending from Llanbabo to Llanrhyddlad. From the summit of Llanrhyddlad mountain towards Carnel's point, a coarse greywacké occurs (219, 220.), intermixed with patches of conglomerate containing rolled pebbles. At the junction between this and the chlorite schist, on the shore to the West of Monachdy, there are irregular patches and stripes of greywacké breccia (210.) embedded in the fine black slate, and not conforming to the direction of the laminar tendency, which appears to indicate a complete intermixture of the materials at their first deposition, and to shew that the laminæ do not mark any order of superposition.

The Western summit of Llaneilian mountain is a decided greywacké (212—214.); very similar in character to that on the summit of Snowdon, in which the impressions of a bivalve shell occur.

A fine grained black clay slate is found on the shore S.W. of Llanfaelog, and in the new road to Holyhead at the nearest point to Llanfihangel church. The small strip which runs from G. 2. Bryngole towards the S.W. is of the same nature.

There is a good exposition of the junction of the greywacké and chlorite schist, between Llaneilian mountain and the point. The line of junction may be traced on the horizontal section formed by the beach, and thence vertically up the face of the cliff. The contact is between a fine grained glossy black clay slate of the greywacké series, and a green slate of the chloritic. The laminæ of each dip towards the N.W., and their union presents a most decided example of a fault. Proceeding Eastwards along the cliff, we come to the coarse greywacke already alluded to. The termination of this is distinct, and it is succeeded by fine green slate, which reposes unconformably upon a black clay slate in the manner represented Pl. XIX. Sect. A. This section is here referred to merely for the purpose of exhibiting the nature of the connection between the chloritic slate and greywacké, it will be again alluded to, and an explanation attempted of the phenomena which it presents. The junction of the greywacké and green slate in the middle of the mountain forms an undulating line down the face of the cliff nearly conformable to the direction of the laminar tendency. The transition from one to the other is gradual, the upper bed of green slate containing a few fragments of a rock resembling the hornstone found between the Paris and Llaneilian mountains, fragments of which are also found in the lower beds of the greywacké.

The laminar tendency of this series is universally inclined at a very high angle to the horizon.

In the central district, on which Llanerchymedd is situate, G. 1.

the bearing of the greatest portion is from the E. of N to the
W. of S., and the dip towards the N. of W   The laminæ are
frequently vertical, often much shattered, and very thin.   Where-
ever the chlorite schist is exposed, along its Western boundary,
it is found presenting the abrupt edges of its laminæ towards the
greywacké.   It is therefore most probable, that the fault exhibited on the coast, between Llaneilian mountain and the point,
is carried directly across the Island.

The principal exceptions to the general direction of the dip
are about Dulas.   In the harbour, the laminæ, though much
confused, dip nearly South, varying to points both to the East
and West.   In this case, therefore, they appear to dip from the
high point of granite on the Llaneilian mountain.

From Llanrhyddlad (on the Western coast) to the Paris
mountain, the average bearing is more nearly East and West
than in the former case, the dip still towards the North.   The
cliff formed by this schist to the North of Carnel's point, presents
the greatest degree of confusion and disrupture among the laminæ,
Pl. XX. Sect. N.   To the South their dip is by no means regular,
but inclines in different directions to the horizon, always however at a very high angle.   Around the point, and again on the
shore to the West of Llanrhyddlad, it assumes a yellow decomposing aspect.

From Llanbabo to the South of Llanrhyddlad the appearances along the Northern line of junction are similar to those
exhibited between Llaneilian and Llanfihangel.   The actual
junction on the coast near Monachdy is obscured by a mass
of diluvium, but judging from the direction of the laminæ on
the horizontal section formed by the shore, the greywacké is
unconformable to the chlorite schist, and therefore presents a
repetition of the facts exhibited on the Western side of Llaneilian mountain.   A few yards to the West of this junction,

the greywacke passes to an unlaminated hard rock mottled with patches and veins of white quartz, and finally assumes a green flinty character similar to that of the chlorite schist at the junction. There is a small cavern in the cliff at this point, the roof and Eastern side of which is formed of the flinty portion, but the schist is again found on its Western side. The union of the two is very evident, the flinty mass reposing upon an inclined plane of the greywacké, Pl. XVI. Fig. 7. The cavern does not resemble a hollow excavated by the action of the sea, but appears as though the upper part had been bent from the Eastern side, when in a soft state, so as to form an arch. It may probably be referred to the nature of a fault, but this explanation admits of difficulty.

On the Western side of the Llydiart mountain, there is a G.4. black shattery clay slate, the laminæ dip from the mountain at a high angle. In the road near Pentraeth their junction has been cut through to the depth of five or six feet, where they meet vertically, and each rock is broken and confused. Following their line of junction towards the North, the mica slate is seen, near Red-wharf bay, to rise from under the clay slate, and presents a smooth rounded surface without any laminar tendency. Between this and the fine grained clay slate, is a thin bed composed of small angular fragments of slate (217.) and at the actual junction it also abounds in small fragments of quartz (218.) loosely cemented together. By one hypothesis this would be called the abraded portion of the two rocks produced by the upheaving of the mica slate. The greywacké may be traced as far South as Llanfihangel.

A confused patch of shattery clay slate, intermixed with G.5. greywacké, is interposed between the chlorite schist and mountain lime, to the East of Llandonna. It does not attain to so great an elevation as either of the formations between which it

is situate, so that it is completely concealed at the spot where
the mountain lime sets on (Pl. XX. Sect. P.)   In immediate contact the chlorite schist consists of a confused talcose rock.
Hardened veins of clay slate intermix with it (234—237.)

3.   The greywacké district placed to the West of the coal-
measures, from Llangefni to the South of Bodorgon, possesses
a different character from the rest (238—250.)   That a portion of
it consists of greywacke, is evident; but whether it belongs to
the present series or to the last, or whether it be not rather a
confused intermixture of both, I did not fully ascertain.

From Llangefni to Aberfraw, repeated instances of greywacké occur, to the East of the schist containing jasper, and
interposed between this and the coal measures.   The schist near
the coal-measures presents its abrupt edges to them, but no
actual appearance of stratification can be traced, and the indications which exist of a laminar tendency are of a very partial nature.
On the N.W. of Llangefni, there is a green talcose clay slate
(242.), occasionally enclosing embedded fragments (243.) and
scales of mica (244.)   It possesses an imperfectly laminar tendency dipping to some point towards the West.   Along its
Eastern termination from hence towards the North, it assumes
a hardened unlaminated character.   At Llangefni it passes to a
green crystalline quartz rock (245—247.), which possesses faint,
but undoubted, traces of globular concretions cemented in a paste
of quartz.   An occasional fragment of uncrystallized matter is
also found embedded.

G. 1.   Through the centre of the Paris mountain, and in the direction of its ridge, there runs a bed of grey cherty stone (252,
253.) cutting through the schist partly hardened (254.), and the
rest assuming a yellow decomposing aspect (255.), full of bleb
and drusy cavities, which also occur in the chert.

The simple minerals found in the extensive and well known

copper mines situate in this mountain, are sulphurets of iron, copper (256.), and lead.—Sulphate of barytes.—Native copper in small quantities (257.), and still more rarely the sulphate of lead.

There are two patches laid down towards the East of the Map, as included in this formation. Certain points of resemblance to portions of the districts already described, seem to stamp them as members of the greywacké series. But they are found under such peculiar circumstances, that it is impossible to speak decidedly on this point.

The small patch to the South of Beaumaris is seen near the G.6. top of Red-hill, and in Lord Bulkeley's grounds, on the slope of the hill above the ferry-house. It appears to be an unstratified mass sticking upon the steep side of the chlorite schist, which rises very abruptly from hence towards Llandonna. It consists of small angular fragments and nodules of clay slate, highly pellucid quartz, and crystallized felspar, either firmly cemented together (263, 264.), or embedded in a hardened argillaceous paste (265—267). The fracture sharp, and approaching the conchoidal. A perfectly flinty slate (268, 269.), with an irregular fracture is associated with it. The specimens bear a close resemblance to those procured between the clay slate and mica slate on the N.W. slope of the Llydiart mountain (216—218.) ; differing from them only in a greater degree of compactness.

The patch which extends from Garth-ferry; about one mile, G.7. towards Cadnant, scarcely reaches above high water mark. The chlorite schist rises abruptly on the West, and forms a high ridge of rugged rocks. The rock at the base is formed of small angular fragments of quartz (270—273.) running together and passing to a compact mass, interspersed with specks of earthy felspar, and fragments of slate (274.). This is intermixed with a few irregular patches of black clay slate, and a compact mica-cious sandstone resembling those in the greywacké about Dulas. (275.) There is scarcely any trace resembling stratification, but the whole rises confusedly towards the chlorite schist.

The greywacké on the opposite coast, immediately South of Bangor, in contact with the coal-measures, seems at first sight to consist of large and small rolled pebbles, firmly embedded in a basis composed of fragments of felspar, quartz and clay slate (280—284.). Upon examination, the rude breccia thus formed is found to possess certain peculiarities of structure, which appear to throw some light upon the nature of a substance found in connection with it, and mentioned in Mr. Greenough's Geological Map of England as " a remarkable steatitic rock, associated with the old red sandstone between Cærnarvon and Conway" (285, 286.)

Many of the pebbles, or rather nodules, are found to indent the surface of a contiguous nodule, as though the latter had been in a soft state, and pressed by the former. The surface of one bed, from which a nodule has been removed, is often abruptly intersected by the surface of another. The surface of the nodules are found to be impressed by the angular projections of the fragments which form its matrix. All the specimens exhibit these facts, and on the natural fracture of one of them (280.), where several contiguous nodules are cut through, they are particularly striking.

These facts admit of explanation, ' by supposing that the nodules are in reality rolled pebbles, which have been softened in some degree, and pressed, since they were brought together. There are, however, other circumstances which appear to destroy such an hypothesis.

The nodules themselves are found in several instances (281.) to be composed of angular, crystalline fragments, which are often sufficiently apparent towards the surface, but which form a compact and homogeneous flinty mass towards the centre, resembling hornstone, and occasionally containing small pieces of pellucid crystalline quartz. Others are wholly formed of quartz, in different states of crystallization, or are slightly intermixed with compact felspar. The matrix also assumes the same characters.

The bed and surface of each nodule, upon a recent fracture, is coated with a ferruginous or ochreous crust. This crust appears also in irregular patches dispersed through the matrix.

From these facts I am inclined to think, that the appearance of a breccia arises from a concretionary structure impressed upon the same kind of fragmental quartz rock, (intermixed with slate) as that which is found on the opposite shore, and that the steatitic rock to which allusion has been made, is a further result of a similar action. This rock consists of white quartz, partly crystalline and partly compact, formed into irregular nodules which run together, but leave several interstices between them filled with a light green talcose substance (285.) The irregular seams which produce the nodular structure are also talcose and ferruginous. Some of the nodules, especially the more crystalline, which attain to one or two inches in diameter, are distinctly composed of irregularly concentric layers (286.). The surfaces of several of these layers are also partially coated with the talcose ingredient, which on weathering becomes detached, and leaves a hollow space between the laminæ.

The homogeneous character which the whole rock must once have possessed, is evident from the numerous veins of quartz or chlorite, which traverse it, always passing through the nodules, however small, which they happen to encounter in their course (283, 284.)

There are veins of crystalline quartz with patches composed of small fragments embedded in them (287.)

How far this rock extends to the South, I did not examine;

but between this place and Cærnarvon, the rocks to the East
of the coal-measures rise high and abrupt.   A specimen (288.)
from them, at Moel-y-don ferry, consists of a flinty mass filled
with embedded fragments of crystallized felspar and quartz,
resembling the internal structure of some of the concretions
just described.   It is traversed by numerous fissures, which
separate it into small fragments, and these also are coated with
the same ferruginous crust as the nodules.

As a concretionary structure was not suspected during the
investigation of these rocks, it is most probable that specimens
might be selected which would better illustrate the facts of the
case, than those which were procured under a different impression.

From Garth ferry (on the Bangor side) as far as Aber, the
dark clay slate is sufficiently regular.   Immediately to the South
of the ferry, it reposes upon a confused mixture of hardened
clay slate of various shades (276—279.), which terminates in the
nodular rock, just described.   It should seem then, that this is
the lowest portion of the greywacké series; but the junction
with the dark clay slate is on too small a scale to enable us to
speak decidedly, though, as far as it is visible, the fact of super
position is sufficiently evident.

Old Red Sandstone.

{Nos. 289 to 372.}

This formation varies considerably in mineral character.   It
occurs as a fine red sandstone, (315—321.), along a narrow strip
o.s. about half a mile in width, stretching S.W. from Dulas harbour
as far as Bryngole.   Even here it is intermixed with shades of
green (313, 314.), and beds of a coarser description (290, 291.)
(295—298.)   A few other small patches of a similar sandstone
are met with in other parts of Anglesea, but the more common

form is that of a coarse breccia. Between Llanfihangel and 0.1.
Llanfaelog it is generally composed of angular fragments of
slate, intermixed with quartz (289.); a character which prevails
as far North as Gwindu. From hence, to Llanerchymedd a
coarser variety is found with pebbles (366.), which on the beach
S. W. of Dulas harbour, form a breccia of the rudest description 0.5.
(293.) The upper beds extend from Bodafon to the mountain-
lime on its East, and consist of a coarse grit, not to be distinguished from some grits of the coal-measures (299, 300.)
About one mile and a half to the South of Bodafon, and a little 0.6.
to the East of an extensive marsh in that neighbourhood, this
grit reappears for a short space, rising through the limestone
which dips from it in opposite directions.

The fine red sandstone round Bodafon mountain, contains 0.5.
small nodular concretions of carbonate of lime (320, 321.)

The strata generally bear in the same direction as the laminar tendency of the last formation, but their average dip is
not so considerable. In the largest district, there is no section 0.1.
sufficiently extensive, which might enable us to ascertain their
nature. In several small quarries about Llechynfarwy, we meet
with a laminar tendency, often thin slaty (308.), inclined at an
angle of 65° towards a point 30° to the W of N This
direction of the dip prevails throughout the remainder of the
district. Numerous edges of broad strata, nearly vertical, project
between Llanfaelog lake and Ceirchiog, and generally possess a
slight degree of curvature towards the S.E., which gives them
the appearance of having been the bases of arches gone to decay
In the greater portion of this district, the subsoil is completely
choked with large fragments of the strata, and as the black
clay slate is found on the shore to the South of Llanfaelog lake,
it is not improbable that the whole consists of a rapid succession
of faults, which have completely dislocated the old red sand-

stone, and left but few patches which may truly be said to remain in situ.

o. 5   In the portion between Dulas harbour and Bryngole, the dip is more gradual and reversed, being about 10° towards a point 20° to the N. of E.   The strata consist of broad, ill defined beds.   In descending the hill to the North of Bodafon towards Dulas, the succession is—a thick bed of green and red sandstone —thin shaly red sandstone—thin beds of green sandstone, with coarse fragments of quartz and slate, and intermixed with partial beds of finer materials—and at the bottom of the valley, the stream to the South of the bridge runs over a shattery black clay slate, the laminæ much confused, but dipping upon the whole at a high angle, in a direction opposite to those of the sandstone strata: an additional reason for supposing these laminæ to be wholly independent of the original order of deposition, and perhaps also for suspecting that the thin slaty beds, mentioned o. 1. in the quarries about Llechynfarway, may be of a similar description.

o. 3.   In the small isolated patch to the N.E. of Llanerchymedd, the strata dip to a point 30° to the W. of N., and are inter-stratified with thin seams of black clay slate: which appears to indicate a gradual transition from the greywacké to the sand-stone.   The termination of the strata to the East is remarkably abrupt, and forms the summit of a low ridge running to the N.E. They repose upon a rotten greywacké, confused and of a yellowish brown aspect (251.)

o. 4.   In the patch to the S.E. of the Paris mountain, the strata dip 50°, and run from the N. of W. to the S. of E., intersecting the former direction at a considerable angle.

It seems highly probable, that no marked separation exists between the greywacké and the old red sandstone, but that the latter merely presents an extreme case of one common formation.

The greater part of this series appears to have undergone considerable alteration since its deposition. This is particularly o. ı. the case about Llanfaelog-lake, Llanfihangel, and in the outlying masses round Llanerchymedd (322—350.). By this change both the coarse (322—330.) and fine grained (331—335.), (341, 342.) varieties assume a more compact texture, arising from an intimate union, and greater degree of crystallization, of the several ingredients. In the coarser specimens, there are traces of large pebbles and fragments (322—326.), some of which may still be detached (327.); but others have become a crystalline mass (328, 329.) passing into the body of the rock, which assumes a more uniform aspect. Towards Llanerchymedd, where the quartzose fragments predominate, the rock in some places passes to a nearly homogeneous mass of quartz (341.)

Bodafon mountain affords a remarkable instance of this o. 5. nature. Without minute investigation, it might be mistaken for an unstratified mass of quartz rock, rising abruptly through the old red sandstone. It is cleft by vertical fissures, breaks into rude shapeless blocks, and presents a barren shattered aspect, not unlike the quartz rock of Holyhead mountain. The summit is subdivided into small elevations, some of which are perfectly rounded and smooth, whilst others are as much the reverse, jagged and splintery. The sides of the fissures which traverse the quartz, are coated with red oxide of iron (345.) The quartz has a flinty semi-crystalline aspect, and is of different shades of red, or mottled with white, grey and green (343—346.) These pass to less homogeneous varieties (347.), in which may be seen distinct traces of a finely granular structure. Others exhibit a coarser texture (349, 350.), and contain quartzose fragments, but so intimately associated with a basis of the same nature, that they cannot always be detected upon a recent fracture. On a weathered surface, however, they are left projecting, owing to the removal of a portion of the matrix in which they are embedded. Small patches of fine white pulverulent matter, probably silica, (for it is neither fusible nor soluble in acid) are dispersed through the solid rock (348.)

This is the oldest formation in Anglesea, in which I found traces of organized bodies, and these were in three separate localities. At one spot, there are three projecting masses of rock, which rise at a few hundred yards to the South of the tenth milestone from Holyhead to Bangor. It is in the mass furthest from the road, and at its Eastern end, that the fossils are to be seen (351—364.). Another locality is where a small rock protrudes in the centre of a field immediately N.E. of Llechynfarwy church, forming the angle between the roads to Llanerchymedd and Llantrisant (365—371.). The third spot is in a quarry on the Eastern side of the road from Llechynfarwy to Llanerchymedd, and about one mile from the latter (372.)

The appearances alluded to, consist principally of the impressions of bivalve shells. In the two last-mentioned localities, the bed is coarse, and partly composed of nothing but rolled quartz pebbles in a gravelly sand (366.) The only species found here appears to be an anomia. It somewhat resembles the common pecten varius (on a small scale), except that the indentations which cause the winged hinge, are in the present instance wanting. The general size is about half an inch wide, but some reach to a full inch. In the first-mentioned locality, the basis of the rock is a finer grained and more compact green sandstone, with partial traces of a slaty structure. Besides the shell already mentioned, it contains the impressions of some other species, which are not in general so well preserved as the former specimens. Among them is one somewhat similar to the last, but the striæ are finer and much more numerous Another is a smooth elliptical shell (352, 353.) A little of the shelly matter still incrusts some of the casts (354.), which are in general coated over with small scales of mica.

Mountain-Limestone and Coal-Measures.

{Nos. 373 to 421.}

A distinction is made, in the colouring of the map, between the mountain-limestone and the coal-measures, although each is supposed to be a member of the same formation. The term coal-measures is meant to include the upper portion, which consists of grit, sandstone, shale and limestone, interstratified with each other, and occasionally containing subordinate beds of coal.

In the most Westerly and principal district of this forma- M. 1. tion, no attempt was made to investigate the exact boundary between the two subdivisions, which would have required more time than a subject of such comparative unimportance seemed to merit. All that is intended in the map, is to note their general limits, and by this means mark their relative position to each other.

The lowest portion of this series consists of a thick bed of stratified limestone, generally of a compact texture, and dark grey colour (373, 374.). It varies also through different shades of brown (376—379.) Sometimes it is composed of a mass of broken fossils firmly cemented together (380, 381.), each of which being formed of calcareous spar, the specimen often assumes a crystalline appearance (382.) Magnesian beds occur subordinate to these. In Priestholme island they are composed of pearl M. 3. spar intermixing with the common limestone (384, 385.)

Chert and chalcedony are also embedded in the limestone, even in the lower strata, before the grit sets on. Some large madrepores (419, 420.) from Priestholme Island, are partly composed of dark limestone, and partly of translucent bluish chalcedony, passing to chert. In some places the cellular coating is chalcedony, and the interior, which was limestone, has disappeared.

Black and white chert, intermixed with the limestone (386.), is more abundant after the grit has made its appearance, and quartzose pebbles are occasionally intermixed with the strata (387.) The limestone becomes more argillaceous (388, 389.) and slaty, and finally interstratifies with clay shale (390, 391.), coarse grit (393—396.) and sandstone (398, 399.)

M. 1. Specks of coal are dispersed through the grit and sandstone, in Red-wharf bay, North of Llanffinnan, and at the Menai bridge (400—403.). To the West of Llanfihangel East, and also to the East of Trefdraeth, coal is worked. At the former place I was informed by the overseer of the works, that it is found in three strata, the thickest of which is two yards, the next one yard and three quarters, and the last four feet. It is peculiarly glistening, and does not contain organic remains. In some clay shale from the pit I observed an impression resembling a flag-leaf.

In the limestone, on the Western side of Red-wharf bay, there are large cylindrical holes filled with grit which formed a portion of the superincumbent stratum, and which is probably the lowest bed of the coal-measures. The partial removal of this stratum has exposed the top of these cylinders, in several of which the action of the sea has worn away the outer crust of the grit, and the hollow in the limestone presents a smooth surface. This circumstance appears analogous to what occurs so frequently in chalk countries, where holes of this description are filled with gravel and sand.

Upon this grit is imposed a bed of shale four feet thick, from which the sulphates of iron and alumina effloresce. This is succeeded by a thick bed of grey limestone, traversed by nodules of jet black and white chert, and filled with the impressions of shells—then a brown sandstone 15 inches—clay shale 4 inches—grit three feet and a quarter—clay shale three inches—brown sandstone 14 inches—dark impure argillaceous limestone to the top of the cliff. This enumeration will serve to shew the nature of the alternations which take place among the strata of this formation.

In the grit, immediately South of the Menai bridge on the M. 2. Cærnarvonshire side, there is a peculiar rock, which appears to form a vertical vein, but the ground is too much covered up to ascertain the point. The basis consists of a coarse heavy red sandstone, containing fragments of quartz, which are also coloured deep red, and through the mass are dispersed numerous small round and oval nodules, from the size of a linseed to that of a small pea (404.) These nodules are composed of concentric crusts of a yellow and brown earthy matter, and exhibit a smooth surface coated with the red oxide of iron. They probably result from some action similar to that which produced the steatitic rock before-mentioned, and which is found at no great distance from this spot.

To the North of Bodorgan, at the Southern extremity of the M. 1. largest district, the grit is composed of small angular fragments of quartz studded with white earthy specks of carbonate of lime (398.) These specks are frequently arranged in parallel lines, inclined to the direction of the strata an effort, if so it may be called, to produce a fissile texture in a coarse substance where it could scarcely have been expected. The particles of the quartzose fragments appear likewise to have undergone a partial re-arrangement; for several contiguous fragments possess a common cleavage. Some of the strata are traversed by fissures, which separate them into blocks, and these decompose in concentric crusts marked by different shades of brown. This compound structure is represented Pl. XVI. Fig. 8.

The only exposed portion of this formation which succeeds the old red sandstone conformably, lies to the E. and N. E. of O. 5. Bodafon mountain. From the Eastern side of Dulas harbour, the mountain limestone stretches towards the S. W. forming a low precipitous cliff, which bounds a marshy ground on its West, as far as Llangefni. From Llangefni to Bodorgan the junction takes place along the side and near the top of an elevated ridge of schist. On the Eastern side of the river, there is another ridge of schist which extends from Caint to Llanddwyn. Between these two ridges there lies a flat swampy ground, beneath which is the only explored coal district of Anglesea.

M. I.   About midway between Pentraeth and Bodafon, the limestone and grit attain to a considerable elevation. The strata are either nearly horizontal, or dip from 5° to 10° to various points between 5° and 40° to the E. of S. Their direction is therefore N. of E., and S. of W. Hence the line of junction from Dulas harbour to Bodorgan must intersect the strata, obliquely to their course, from the lowest upwards in a regular succession.

Between Llangefni and Bodorgan, several opportunities occur of examining the nature of their union with the adjacent rocks. In every case the limestone and grit are confused and broken. The schist also rises in a shattered and abrupt manner, dipping from the limestone wherever there happens to exist any tendency to a laminar structure. In some places it projects in peaks surrounded by the limestone or grit, at others it encloses small patches of the latter. Immediately S. W. of Llangefni there is a quarry of limestone, on the brow of the hill rising from a marshy ground, which presents a section exhibiting the confused nature of this junction, Pl. XVII. Fig. 1. The strata are nearly horizontal, but sensibly bent upwards next the schist on either side. In a small elevation lately cut through in forming the new road from Bangor to Holyhead, on quitting the marshy ground, the limestone and shale dip towards the S. E.

at an angle of 45°, and exhibit a fault by which they are up-heaved towards the schist. Similar appearances to these may be seen in different quarries between this spot and Trefdraeth.

At Bodorgan, an isolated patch of schist rises through the confused and dismembered grit. Between the grit and schist there is a loose breccia chiefly composed of angular fragments of the latter (392.), which may be accounted for in the same manner as the breccia interposed between the clay slate and mica slate on the N.W. side of the Llydiart mountain.

On the W of the Pentraeth river, near Red-wharf bay, the limestone dips 45° to the W. of N. In Pentraeth their inclination reaches as high as 80°. At Caint they are confused, broken, and sometimes contorted without fracture, Pl. XVII. Fig. 2. But on proceeding to the West of these several places, we find the strata nearly horizontal. They present several low cliffs, which are not so abrupt as those on the Western boundary of the series. These facts seem to indicate, that the grit and limestone terminate abruptly against the schist to the East, with the intervention of a few hundred yards of disruptured and broken strata.

East of Llandonna, the limestone presents an abrupt cliff M. 3. to the sea, the strata are nearly horizontal, their edges reposing on an inclined plane, the summit of which is chlorite schist; but at a lower elevation we find the shattery schist before-mentioned, so that the limestone overlaps this in the manner represented Pl. XX. Sect. P. Large fragments of the limestone strata are scattered over the steep sides of the chlorite ridge between this spot and the sands of Red-wharf bay. This district extends to the East as far as Priestholme island. Near the point to the North of Penmon, some coarse grit sets on. The dip is towards the E. of N. at no great angle or inclination.

The appearance presented by Great-Ormes-Head Pl. XVII. Fig. 3. on the opposite coast of Cærnarvonshire is such as might lead us to expect a continuation of the same strata at that place. I did not visit the spot, but it is evident from the opposite coast, that a considerable indentation Northwards takes place towards the Eastern extremity of the Head. This must expose each stratum at some point further to the North than its Southern boundary, where (owing to the former dip towards the E. of N.) it will be seen at a less degree of elevation, which would give rise to the deceptive appearance of bason shaped strata exhibited towards the East of the figure.

M. 2. In the tract lying to the S.W. of Bangor, the strata belong to the coal-measures. A little to the South of the spot where they first appear, there is a large limestone quarry, which lies beneath some beds of grit and shale, and is possibly a portion of the series belonging to the mountain-lime. The strata here also dip towards the E. of S., and are in contact on the East with greywacké, and the older rocks dipping also, when laminated, in nearly the same direction. The line of junction is obscured by a cultivated valley. They are bounded by mica and chlorite slate on the West.

The fossils found in this formation are anomiæ, madrepores, trilobites, and others identical with those from the mountain-lime of England (405—421.)

Magnesian Limestone.

{Nos. 422 to 474.}

M. 2. To the South of Plas-Newydd park, there commences a series of limestone and sandstone strata, which overlie the coal-measures,

---

This place lies to the East of Priestholme island, but is without the limits of the Map.

and appear to belong to a separate formation. They are better exposed, and may be examined with greater convenience on the opposite coast.

The lowest portion consists of rolled fragments of limestone, cemented together by argillaceous and calcareous matter (422.) To this succeed beds of limestone, grit, and sandstone, variegated with deep yellow and brick red colours. Their order from the bottom is,

feet

1. Yellowish brown sandstone (424.) 5

2. Compact and crystalline grey limestone, with specks $\Big\}$ .. 5
   and cavities filled with yellow ochre (423.) ....

3. Bluish shale, (thin bed) ....

4. Compact flinty dark grey sandstone, nearly a pure
   quartz rock, which separates into rude distinct masses
   coated by a deep yellow ochre (444). This inter- $\Big\}$ 10
   mixes with,

5. Fine red, striped sandstone, (442.), containing fragments of broken fossils (436.). ....

The two last beds contain variable portions of lime.

6. Thick bed of compact red limestone (427, 428.), which has
   been quarried to a considerable extent.

Upon this are imposed other strata of a similar nature to those described, whose order of superposition it is not so easy to ascertain. They are all more or less characterised by containing beds of bitter spar (437—440.) The fossils are generally in an imperfect and shattered state, intermixed with pebbles (434, 435.) The more perfect madrepores are frequently traced in deep red upon a light ground (432—434.). These fossils appear to have belonged to the mountain lime, and may be considered as embedded fragments in the present formation. Although I found no good section, by which any positive

information might be obtained of the nature of collocation between this series and the last, still it seems probable, that they
lie unconformably to each other.   Immediately South of Plas-
Coch, the black limestone and shale of the coal-measures dip
towards the E. of S., and a few yards to the East of this spot,
the red beds of grit are found dipping in a contrary direction.
On the opposite coast, to the East, of Plas-Newydd, the lowest
strata of the red grit and limestone dip gently to the E. of S.,
apparently conforming to those of the coal-measures; but suddenly their dip is considerably increased, as if they were reposing upon the brow of a steep hill.   As the red beds appear to
be entirely wanting over the marshy tract in which the coal of
Anglesea is situate, it is not unlikely that in the present place
they overlie a considerable body of that formation.

New Red Sandstone.

{Nos. 475 to 482.}

Over the strata of the last series, there occurs a rude mass
of argillaceous and sandy materials (475.) intermixed with large
fragments derived from the older rocks.   The basis is occasionally consolidated into thin laminæ, giving rise to a slight
appearance of stratification.   The whole is of a deep red colour.
It commences a little South of Moel-y-don ferry, on the Eastern side of the Menai, and extends as far as Cærnarvon, but
in Anglesea it forms only a small hummock on the North of
Tan-y-voel ferry.   Both this and the preceding formation terminate to the E. and W. in the same abrupt manner as the
coal-measures.

Many of the fragments dispersed through it, are of a large
size, and generally consist of quartzose materials.   Some are
of grit in which the fragments run together and pass to a homo-

geneous quartz rock (**476, 477.**), others approach chert (**478.**) or hornstone (**479—481.**), and nearly all are tinged red.

A fault in the coal-pit near Llanfihangel East, contains fragments of quartz intermixed with red sand (**482.**), and may probably have arisen from a portion of this formation having filled up a fissure.

Trap Dykes.

Although these form but inferior members among the unstratified rocks, still it seems advisable to commence this part of the account with their description; since the facts which they present tend materially to confirm the conclusions drawn from phenomena observed in more extensive districts, apparently of similar origin.

As their number is very considerable, and as a detailed description of each would only occasion repetition a selection has been made of those which are accompanied by appearances of the greatest interest.

Where their course is sufficiently exposed, it is represented on the Map in the usual way, by a particular colour placed between parallel lines. But there are many slight indications of trap, where it is either impossible to trace the course of the dyke, from the concealed nature of the ground, or it only presents an isolated mass rising through the schist. In these cases, the locality is marked by placing a (τ) as near the spot as possible. References of this description will seldom guide a second person to the dyke, but may serve a different purpose, and shew the number and relative situation of the places where trap was actually met with.

Mr. Underwood has submitted specimens of several of these dykes to the examination of Professor Cordier, whose method of analysing the basalts, and accurate knowledge of their mineralogical composition, is too well known to need a comment I have enriched this account with his description of several specimens, and it will be seen that, judging merely from their composition and texture, and unacquainted with the geological phenomena by which they are accompanied, he supposes them to be of volcanic origin.

The dyke with which I shall first commence, is seen on the shore, immediately South of Plas-Newydd, between two and M. 2. three hundred paces from the landing place. The phenomena with which it is accompanied are exhibited on a scale sufficiently large, and are besides so unequivocal in their nature, that the results deducible from their examination may be considered as of the greatest importance towards the elucidation of this class of rocks.

The width of the dyke is 134 feet, and it cuts perpendicularly through strata of shale and limestone. The strata on each side form an abrupt cliff, about 15 feet high, but the dyke affords a gradual ascent to the top, arising from the effects of its decomposition. On the beach also the same cause has contributed to produce a slight excavation worn by the action of tide. In fact the decomposition is generally so far advanced, as to render it capable of being dug with a spade, and it is applied to the same purposes as coarse sand, being mixed with mortar as a material for building.

There is no absolute certainty of its further extent Westward, than about forty or fifty feet, through which it may be traced in Plas-Newydd Park, where some person, perceiving its nature to be different from that of the surrounding strata, has been at the pains of driving a level up it, in the fruitless hopes of discovering a metallic ore.

The substance of this dyke is "indubitable basalt, composed of felspar and pyroxene." *Prof. Cordier.*

The effect of decomposition probably does not extend to any great depth, for the workmen soon find the rock become too solid to be used for sand. In the more solid parts the texture is rather coarse, and the proportion between the felspar and pyroxene variable. Sometimes these are nearly equal (483.), but in general the latter predominates considerably (484—486.), and the rock is of a dark colour Sometimes the basalt is very compact (486.), and does not exhibit any signs of decomposition. Carbonate of lime is very generally disseminated through every part. In the midst of the friable and decomposing portion, there occur irregular, concretionary nodules (487.), about the size of walnuts, which consist principally of felspar rudely crystallized in one mass, through which are dispersed small crystals of black pyroxene. These nodules are remarkably tough.

The dyke is found on the opposite shore of the Menai, but there is no section by which an opportunity might be afforded of obtaining any satisfactory conclusions. Similar varieties of basalt occur in five other places, which bear so nearly in a straight line with the two already mentioned, that it is highly probable they are all portions of this dyke, but the concealed nature of the ground renders it impossible to obtain certain information of the fact. The nearest spot is in a quarry, on the North side of the road which runs South of Plas-Gwyn to Llanddaniel, and at 170 paces East of the bridge over the Braint river.

Two other places, within fifty yards of each other, were lately exposed upon digging the foundation for the new road from Bangor to Holyhead. They lie to the South of Llanfihangel-East, on the brow of the hill to the East of the marshy ground over the coal-measures, and the basalt is continued along the steep Northern bank of a small stream which here crosses

the road, and then runs parallel to it on the South. The last
spot is on the opposite side of this marsh, about a quarter of
a mile to the S.W. of Llanchristiolis, at a spot called Tin-rath.

The cliff which bounds the dyke, at Plas-Newydd, is composed of clay-shale and argillaceous limestone. On the Northern
side it may conveniently be divided into four parts. The lowest
consists of thin, dark, shaly beds, containing a considerable
quantity of lime. In it are found the impressions of small
anomiæ common in this formation (490.) Upon approaching
the dyke, the shale undergoes various degrees of alteration.
At fifteen feet from the contact, it forms a compact bluish
grey mass (492.) with spots of a fainter colour. The substance
of some fossils, which coat a natural cleavage, puts on a crystalline appearance. In contact, it is of a very compact nature ;
bluish-green (493.), with irregular streaks of a lighter tinge,
fracture conchoidal ; not easily scratched. Several small crystalline plates of carbonate of lime are dispersed through the mass.
The shaly structure disappears, in a great measure, upon approaching the dyke, but a partial separation into parallel beds
is still evident.

Owing to the variable nature of this shale, we cannot expect
a gradual passage from its original aspect to the indurated part.
Together with the hardened variety just mentioned are patches
(at five feet from the dyke) of yellow clay, slightly indurated
(494.), and intermixed with fossil shells composed of crystallized
limestone (495.) At two feet, a light yellowish clay forms the
basis, through which are dispersed patches of dark brown ; and
black specks are placed in the centres of small spherical kernels of crystallized carbonate of lime (496.) On a weathered
surface these are removed, and the rock is indented with the
small cavities which contained them.

It is in the next portion, above this, that the more striking phenomena, on this side of the dyke occur. At fifty feet from the dyke it consists of a soft, dark coloured plastic clay shale, which separates into thin laminæ (498, 499.) On approaching the dyke this becomes, at thirty-five feet from the contact, rather indurated (500.) At ten feet, it forms a cherty mass (501.), with a splintery fracture, and of a buff colour, associated with patches and streaks resembling black flint (502). In it are patches of highly crystalline limestone. It scarcely admits of being scratched by the knife.

In contact, it is a hard porcellanous jasper, which readily cuts glass, extremely splintery, and the fragments fly from the hammer in all directions, producing an appearance similar to the effect of fracture on unanealed glass (503—509.). Its colours are light and dark grey, sometimes intermixed in irregular stripes parallel to the former position of the shaly structure. Sometimes the light grey assumes a reddish tinge (505.), and the specimen somewhat resembles a piece of fine porcelain, and is translucent at the edges. The fracture is splintery-conchoidal. Another variety is dull greenish-brown, and more nearly resembles a piece of chert (504.)

The impressions of broken shells, generally anomiæ both large and small, occur in the interior of the solid mass (507, 508.) It does not divide into parallel beds except in one or two instances, where the natural divisions are formed by a crust containing impressions of these anomiæ. This crust has a feruginous aspect externally, but possesses a resinous fracture and lustre within, and effervesces in acids.

There is a circumstance which generally takes place on the surfaces of small natural fractures or flaws, dispersed through hardened shale in contact with this and other dykes in Anglesea. On such surfaces there are small glittering plates (503.) from less than $\frac{1}{20}$ of an inch to full $\frac{1}{4}$ of an inch squared.

Different plates may be exhibited by presenting a surface at different angles of inclination to a ray of light, so that it seems to be entirely composed of them. These plates cannot consist of one uniform façet, for in that case the fracture would present an uneven surface of re-entering angles. Each plate is therefore formed by reflection from an aggregation of several minute polished facets, inclined at a common angle to the surface in the same plate; but this angle in different plates must vary.

The third division of this cliff consists of a dark argillaceous limestone about three feet thick, containing impressions of shells (488.). This is also capable of a partial subdivision into thin laminæ, but does not possess so decidedly a shaly structure as the two last, neither is the proportion of argil so great. In contact it forms a remarkably tough, close-grained mass of a speckled dull green and brown colour (489.)

A similar argillaceous limestone in contact with the dyke, on the opposite coast, is not so much changed. It assumes a hard character without much alteration of colour, and becomes finely granular and crystalline. Through it are dispersed specks of pyrites and impressions of shells (497.)

Above this we find another bed of clay-shale (510.), whose contact with the dyke is not exposed on the Northern side It constitutes the main body of the low cliff to the South. This shale is also partially converted to a flinty mass similar to that already described. The flinty portions lie in irregular strings of various thickness, parallel to the position of the beds, and the rest of the shale assumes a confused appearance of crystallization and globular structure. The perfect crystals which occur intermixed with the mass, present two decided mineral varieties. A description of the specimens, selected for the Woodwardian Museum, will perhaps be the best method of enabling others to judge of their nature and mode of formation.

(511). Shale, of the consistence of hard clay, passing on one side to a globular structure, of a dirty white, earthy aspect. The globules, from one-tenth to three-tenths of an inch in diameter, run into each other, and are harder than the rest of the mass. It effervesces rather briskly for some time, without falling to pieces.

(512). The whole possesses a concretionary structure. The concretions white, some of the same size, but harder than in the last specimen. These are interspersed with smaller globules, about the size of mustard seeds. The interstices are filled with a soft brown clay. A partial effervescence may be obtained from the mass, but a detached globule affords none. Several of the concretions present crystalline facets of a trapezoidal form.

(513). With irregular concretions three-fourths of an inch in diameter. Upon fracture these are found to consist of an outer crust of more crystalline character, and lighter colour than the interior. The surface of each globule is very rough. Smaller globules occur within the larger.

(514). Several perfect crystals with twenty-four trapezoidal faces (the common form of leucite and analcime) are scattered through this mass, and traces of a crystalline structure are evident in the compact portions. The more friable parts, between the crystals, consist of a white earthy substance which effervesces briskly, but leaves a considerable residue.

(515) Globules and imperfect crystals closely united, and occasionally presenting a compact mass, resembling specimens of the hardened shale. In this specimen there occurs the impression of a valve of an anomia, two inches in width. The crystals are studded over, and pass through it, and small fragments, detached from the shell, lie embedded near it.

(516). Thin plate of hardened shale, striped light and dark grey, studded as in the last. The brown clay is left filling up the interstices between the crystals. Impressions of shells on each side of the specimen.

The original slaty structure of the shale modifies the shape of all these specimens, and the more perfect crystals lie on the sides which were parallel to the position of the beds.

I have great pleasure in adding to this account, an accurate examination, and a minute analysis of the perfectly formed crystals, kindly undertaken by Professor Cumming.

" They are slightly electric by friction; they scratch glass; they readily vitrify before the blow-pipe, and gelatinize with acids. The specific gravity of some detached crystals was 2.293; that of a mass 2.394. By exposure to a red heat there was a loss of 5 per cent."

" The mineral was examined in the usual method, by repeated digestion with muriatic acid, the residue being fused and boiled with caustic potash. The muriatic solution, dried and heated to redness, to decompose the muriates of iron and alumina, was dissolved in water; a small quantity of lime was separated from the solution, which by evaporation gave crystals of common salt. Silex was obtained from the alkaline solution by muriatic acid; alumina and iron were precipitated by caustic ammonia, and lime by the oxalate. The iron and alumina were separated by boiling with caustic potash, from which the alumina was precipitated by muriate of ammonia. The results were,

$$
\left.\begin{array}{lr}
\text{Silex} & 49 \\
\text{Alumina} & 17 \\
\text{Lime} & 12 \\
\text{Iron} & 4 \\
\text{Soda} & 9 \\
\text{Water} & 5 \\
\text{Loss} & 4
\end{array}\right\} 100
$$

' Hence, the crystals appear to be analcime with excess of iron."

Small crystals are found embedded in the soft earthy matter, placed between the globules and the large crystals. They are sometimes beautifully symmetrical, but more frequently their facets are less regularly disposed Their colour is dull opake white.

The remaining specimens present a crystalline substance distinct from the last, but the general character and mode of aggregation is nearly similar. The crystals here assume the form of the rhomboidal dodecahedron, and are generally found in greater perfection than the former, most of the concretions possessing crystalline faces. They are often pressed together into a confused mass, with somewhat the appearance of Coccolite.

(517). Several good crystals; the largest diameter of the rhombic faces is two-fifths of an inch. These are attached to the hardened shale, which contains thin seams of crystallized limestone.

(518). Mass of the crystallized matter; the diameter of the largest rhomb three-fifths of an inch The interior of some of the large crystals is made up of minute crystals and globules. A thin coat of carbonate of lime covers some of the faces. The edges of some of the dodecahedrons truncated.

(519). The facets exhibit a satiny lustre; the intervals between the crystals are filled with a dark plastic clay and crystallized carbonate of lime. The internal structure of some crystals presents a series of dodecahedrons, one within the other.

(520). Very minute crystals fill up the intervals between the larger, and scarcely any clay is present.

(521). Some large crystals of the dodecahedron with truncated edges, seven-tenths of an inch in diameter. Facets well defined, possessing a resinous lustre and olive brown colour. These are embedded in a mass of crystalline matter of the same kind, and the whole attached to some hardened shale.

(522.) Crystals on hardened shale. This shale presents a mottled appearance, as though numerous globules, running together, were firmly cemented by a substance of similar nature but different colour.

(523). Hardened grey shale with white globules, more perfect than in the last, embedded but not distinct from the mass, so that the fracture presents white circular patches on a dark ground.

Professor Cumming has examined also the nature of these crystals "but they appeared so decidedly to possess all the characters of the garnet, that it was unnecessary to enter upon a minute analysis. They readily scratch the former crystals; their specific gravity is 3.353; they do not easily fuse before the blow-pipe "

c. 5.　　　At Cadnant, there is another large dyke of a similar description with the last.

On the opposite shore of the Menai, it cuts through the limestone and shale. The effects which it there produces on the surrounding strata are not so powerful as in the former instance, but the marks of violence by which it is accompanied are more

---

* I am fortunate in being able to add another instance where garnets have been found under similar circumstances of association. Sowerby, in his British Mineralogy, Vol. II. p. 37. Tab. 120, mentions minute garnets, received from the Rev. J. Harriman, "the crystals from the size of a small pin's head to an extreme degree of minuteness—of the form of the rhomboidal dodecahedron,—mixed with a rough mass of their own nature, which seems to incorporate with some quartz—the matrix of carbonate of lime and a siliceous substance, resembling dull reddish jasper." He neither alludes to their locality, nor to the circumstances under which they were found. Whilst Professor Sedgwick was in the North, during last summer (1821), I wrote him an account of the discovery of the garnets and analcimes at Plas-Newydd. Shortly afterwards, he accidentally heard that garnets had been found, in the Mountain limestone formation, in High-Teesdale, and imagining that their geological relations might be the same, he (directed by Mr. Harriman) visited their locality, and though he did not procure any garnets, found the exact matrix described by Sowerby. This proved to be some altered shale and limestone in contact with a large overlying mass of basalt about half a mile below Caldron-Stuno, on the Northern side of the stream. The fact therefore is doubtless of a similar nature to that above-mentioned, though the crystals are not so well exhibited, and are confined to one mineral species.

striking   Considerable masses of the stratified rock lie completely embedded in the trap, and some portions are partly envelloped by the dyke, which ramifies and passes round them.

The composition is "most certainly dolerite; felspar and pyroxene". *Prof. Cordier.*

The main body approaches to the same earthy state as the dyke at Plas-Newydd, and through it are dispersed patches of a more compact nature (524.). This structure is common to all the dykes of a similar character in Anglesea. The harder portions resemble irregular spheres, pressed together, which decompose in concentric crusts Sometimes these hard portions run into small columns which break into separate joints (527—529) In some parts we find scales of dark brown mica (525), a circumstance of rare occurrence in the dykes of Anglesea. Carbonate of lime, in small quantities, is dispersed through the rock. In contact with the surrounding strata, to the South, it passes to a dull earthy wacké which separates into prismatic masses (526.). In one spot, on the Southern side and near the top of the cliff, it assumes a peculiar character, consisting principally of crystals of jet black pyroxene with a little felspar and lime intermixed, and through it are dispersed nodules of mesotype with crystallized carbonate of lime.—This portion presents concretionary masses about one foot and an half cubed, and is in contact with an embedded mass of the stratified rock The natural fissures are coated over with a crust of carbonate of lime, formed into an irregular aggregation of rhombs.

The mesotype is also found in the earthy portion of the dyke, but the radii composing each nodule separate upon removing the specimen.

The entangled strata, just alluded to, form a flinty mass (532.) coated and penetrated by carbonate of lime in a pulverulent state (533.). Other portions consist of hardened sandstone

(534, 535.) intermixed with crystalline limestone (536.), and cherty masses, where the natural fractures are covered with the peculiar glistening plates, formed from an aggregation of small parallel facets, in the manner already described (537.). Small crystals of selenite are found in the fissures of the same specimens.

As might be expected, the effects produced upon the surrounding strata are not so marked in this place as at Plas-Newydd, where the dimensions of the dyke are larger. The alterations, however, which they sustain, are sufficiently striking On the Southern side, the shale and limestone assume a ferruginous, or ochreous appearance, and the passage of the clay to a flinty character (538.), and of the compact limestone to a crystalline (539.), together with the intermixture of the two (540.), where the stratum is of a compound nature, are quite distinct. In one bed of clay shale, which partly assumes the flinty appearance (542.), there are some portions contiguous to the dyke which are found in an earthy form. They are however in a state of decomposition and consist principally of yellow ochre (542.). The casts of large madrepores, partly composed of chert, and partly of limestone, found in the same stratum, (544.), become distinctly crystalline upon approaching the dyke (543.).

On the Western shore of the Menai, the whole width of this dyke is not exposed ; it may be traced for 78 feet, and is seen, in the direction of its course, along the steep Southern bank of the stream which empties itself at this spot. It is much more compact than on the opposite coast, a circumstance probably owing to the nature of the surrounding rock ; for none of the dykes which intersect the limestone and shale attain to so great a degree of compactness as the generality of those which are found among the schist.—It is here styled by Professor Cordier, "a true dolerite, having the ingredients,

pyroxene, *fer-titané*, and felspar, well characterised." The proportion which the carbonate of lime bears to these, is very trifling
(545.). Proceeding inland the dyke is soon lost, Pl. XVIII.
Fig. 1. is an ideal section, beyond this, along the direct
course of the dyke, the dotted lines representing the concealed
portions. Parallel to this section there runs a hollow valley,
bounded on the North and South by abrupt elevations of schist,
and it is beneath the Southern boundary that the dyke is supposed
to take its course. On arriving at the top of the hill, three
distinct strata, in the chlorite schist, are seen rising in the contorted manner represented in the figure, and forming a small
elevation on the top of which there are large stones arranged in
circles, which are said to be the remains of an ancient British
town. To the West of this spot the trap again protrudes, by
the side of the road leading from Beaumaris to the post-road,
and a little further to the West it is once more seen rising
through the schist, which is hardened (550.), contorted, and
partly changed to ochre (551.).

The central portions of the dyke bear the same character
as at Cadnant, but towards the outside it assumes a greater
degree of compactness (546, 547.), and in contact, forms a dark
grey, and nearly homogeneous mass, very tough, and with a
conchoidal fracture (548, 549.). Professor Cordier remarks upon
a specimen from this spot, "This is a desideratum; the air-holes,
and greater degree of compactness where in contact with the
schist, render all further discussion as to its igneous origin
perfectly ridiculous." The fact which this dyke presents, of
becoming more uniform as it approaches the surrounding schist,
is repeatedly to be met with in many parts of Anglesea, and
may indeed be said to form a general character in dykes of
this description.

There is a dyke, 40 feet wide, at Moel-y-don ferry to the

South of Plas-Newydd; an horizontal section of which is exposed on the Western shore. It reaches to the opposite coast, rising through a projecting mass of limestone and shale which forms a small promontory to the South of the ferry-house.

Its composition is "felspar and pyroxene. The laminæ composing the crystals of felspar are all in the same direction ; a circumstance similar to what takes place in the lavas of Ætna and Tenerif." *Prof Cordier.*

Parallel to this, on the Western shore, and at twenty-four feet to the South, there is a small string, five or six inches wide, of the same kind of trap ; and the main dyke itself ramifies towards the North. Between the two, the limestone is considerably altered In it are a quantity of madrepores ; the whole assumes a crystalline character, being formed of small plates loosely cemented together, which gives the specimens the appearance of a sandstone. The madrepores are traced by black plates, and the basis in which they are embedded by plates of a light colour (554—558.). Some portions of the madrepores are not crystallized, whilst the intervening limestone is ; which produces an easy separation at the joints, and furnishes a better opportunity of examining the internal structure of these bodies, than could otherwise have been obtained (554, 555.). Possibly the black plates, which trace out the former space occupied by the madrepores, may owe their colour to the carbonization of these bodies.

Through part of this dyke (553.), and in some others in the neighbourhood (559—563.), there are dispersed small globules of a white transparent mineral, the lustre of which resembles that of stilbite. It occasionally possesses a light blue tinge and opalescent aspect. To a slight depth, on the exposed portions of the dyke. all the globules are decomposed to an ochreous powder. Before the blowpipe it turns black, but does not appear to sustain any further change. I suspect this to be a modification of a substance which Mr. Underwood informs me that he found in the dyke nearest to Plas-Coch, on the S.W. (562, 563.). Mr. Underwood says, "What I took for olivene, (and which, though bright green when I broke it), became black in a few days, (an effect which Cordier had once experienced in green earth), and although it has a vitreous fracture, may yet be scratched with a knife." There can be but little doubt that this is the mineral named chlorophæite by Dr. Macculloch in his Western Islands, Vol. I. page 504.

A dyke, composed of little else than crystals of dark green pyroxene coarsely aggregated (564.), and intermixed with pyrites (565.), intersects the strata at the coal-pit on the S.W. of Llanfihangel East.

M.I.

Some dark lead-coloured clay shale (567.), in contact with it, passes to a hard jaspideous mass scarcely differing in colour from the unaltered portion (568.). Other strata of shale, grit, and limestone, sustain alterations similar to those already described. Where the dyke comes in contact with the coal, the latter is converted to a scoriaceous cinder full of air-blebs, and traversed by cracks and flaws, many of which are filled by crystallized carbonate of lime (569.). Upon removing this by an acid, a perfect cinder is exhibited, which will neither inflame nor emit any smoke before the blow-pipe. A rude columnar structure may be traced in some portions of this substance (570.), the prisms being about half an inch in diameter, and this is also visible where the apparent conversion to a cinder is less evident, but where the inflammable matter is equally wanting (571.).

Equiaxe rhombs of crystallized carbonate of lime thickly coat over the natural fissures, both in the dyke (566.) and in the cinder (571.).

Q. 1.   Another dyke extends from the neighbourhood of the South
Stack to the Southern extremity of Holyhead island.   The
average width of this is about sixty feet.   It first appears to
the S.E. of the Stack, forming some dark brown rocks, projecting from the sea, at the base of the cliff.   It then cuts
through a small headland not less than 200 feet in height, and
may be distinctly seen in the cliffs to the East and West, but
no trace of its intermediate course is apparent on the surface;
shewing how very easily a dyke of this description may escape
observation, unless accompanied by a succession of denudations.
It again enters the cliff on the S.W. side of Hen-Borth, and at
this spot presents a picturesque mass of rock, rising abruptly
from the shore, the base of which is surrounded by the sea at
high tide.   It is then lost until, after crossing some high ground,
we arrive near Port-Dafreth, where its course is marked, along
a sunken track lying between two ridges of schist, by several
portions of the trap projecting through the surface.   It is seen
in the cliff on each side of Port-Dafreth, and may distinctly
be traced to the S.E. through the next promontory.   It is then
lost as a continuous body until we arrive about half a mile to
the S.W. of Borth-Anna.   In the intervening space, however,
there are numerous strings of trap rising through the schist, one
of which exhibits a termination upwards, Pl. XVII. Figs. 5, 6.
At the spot where it re-appears to the S.W. of Borth-Anna,
this circumstance is remarkably well exhibited.   It so happens
that there is a vertical section in the very direction of the dyke,
and the trap is seen for ninety feet in length, capped by the
schist from fifteen to thirty feet in thickness, Pl. XVII. Fig. 7.
The surrounding schist is in a most confused and contorted
state, and considerable portions of it are entangled in the trap.
Immediately beyond this spot the dyke regains its original
character, and presents several hard projecting masses, rising

between two ridges of schist, as far as the marsh on the North of Rhoscolyn. It is found in three places, after crossing this marsh, which bear in the same direction as before; after which it is concealed until we arrive at its Southern termination, where it runs out to sea along the Western side of a small bay.

The most interesting phenomena exhibited by this dyke, are the various changes which it assumes in its mineral character. These changes are not merely such as are presented by different portions of it, at considerable distances from each other, and where it is possible that some doubt of their perfect continuity might exist, but are such as may be traced in parts of one mass.

In Port-Dafreth, there is a recent section which enables us to investigate this point with considerable facility. The main body of the dyke assumes a dirty brown earthy aspect, as though it were in a state of semi-decomposition, and through it are dispersed small crystals of felspar and pyroxene (602, 603.) Some portions of this are filled with nodular concretions about the size of peas, composed of crystallized felspar interspersed with small crystals of pyroxene. The earthy base becomes washed out by exposure, and the nodules project upon the surface (604.)

The more compact portions of the dyke resist the action of the weather, and, when the softer parts are decomposed and washed away, present projecting masses of bare rock, which enable us to trace its course with greater facility. The usual character is felspar and black pyroxene, intimately associated, and possessing a tough, uneven fracture. Although the individual crystals of each substance are small, there are numerous traces of a laminated structure, common to several of them, scattered in different directions throughout the specimen (605.). In one spot, at **Port-Dafreth**, this variety is amygdaloidal. containing kernels of white chalcedony (606, 607.).

In the small strings given off among the schist between Port-Dafreth and Borth-Anna, the trap always assumes a more compact character, resembling the varieties found at the termination of the dyke at Cadnant to the West (608, 612—614.)

In the Southern division of Holyhead island, the crystals of felspar and pyroxene are generally larger and more distinct. Professor Cordier considered the specimens from this portion " as belonging to the most ancient dyke in Anglesea. The felspar is whiter, and the pyroxene greener. It perfectly resembles the granitoidal ophites of the Vosges, which in those mountains exists in powerful beds. In the ordinary diorites, the felspar is greener, and the pyroxene blacker. The rock is highly interesting, and merits a very strict investigation." The characters here specified are partial, and the more ordinary varieties of this very dyke have, as he describes, the felspar greener, and the pyroxene blacker (515—520.). Hence it should seem, that the distinctions of age, deduced from mineral character alone, are not applicable to the dykes of this country.

The appearance of alteration impressed upon the rocks in contact with this dyke are not so striking or so general as those afforded by the dykes already described, which intersect the strata of more recent formations. It is possible that many facts of this nature have been obliterated by subsequent decomposition, or may actually exist where we are not sufficiently acquainted with the original character of the rock itself, to be able to determine whether an alteration has taken place or not. It is however evident, that such has been the case in several instances. On the Western side of Port-Dafreth, there is an indentation to the South of the dyke, formed by a re-entering angle in the cliff, and parallel to its course. The thin slaty laminæ of chlorite schist, which project round the sides of this hollow, are remarkably sonorous when struck by the hammer, and consist of compact and rather splintery quartz, translucent, and imperfectly tinged green (622.). The surfaces of the thin laminæ are coated with scales of chlorite (623.).

Where the dyke is contained between parallel walls of the schist, and appears as though it were filling up some large crevice, the effects are never so striking as in those places where it ramifies and becomes intimately associated with the surrounding mass. It is in these cases that the trap assumes a more compact character, especially where fragments of the schist have become entangled in its substance. These fragments entirely lose their original aspect, and present a finely striated and contorted mass, blending with the trap, and forming, as it were, a part of its substance (613.). The line which separates the dyke from the surrounding schist is distinct, and a blow will generally detach them; which renders it difficult to procure a specimen exhibiting their junction.

At the spot where it intersects the serpentine, the dyke ramifies for a short space, but soon re-unites. The mass thus enclosed between the two branches consists of dark argillaceous matter, which shatters into small fragments bounded by most irregular though natural cleavages. Through it are dispersed small crystalline plates (621.).

The dyke which runs, from the North-Eastern side of Holy- Q. 1. head mountain, towards the S.E., to a spot between Port-Dafreth and Borth-Anna, possesses precisely the same characters as the last. At its Northern termination the trap has been removed by the continued action of the sea, and its original walls, composed of quartz rock, form a small bay about eighty feet wide. In this as well as in the last dyke " there is less *fer titané* than is usually found in dolerite, but there is more pyrites." *Prof. Cordier.* (624—627.). The quartz rock in contact is partially altered, and has become charged with a considerable quantity of felspar, in a state of decomposition (630—632.). In other places, where it remains hard, it loses its usual crystalline appearance (633, 634.). In the trap, there are several masses of hardened schist (628.), and also irregular strings or veins of breccia, composed of angular fragments of quartz, felspar and schist (629.).

The evidence for the continuity of this dyke is not so clear as for that of the former. From the top of the cliff may be seen a hollow tract, lying between two ridges of schist, stretching in a South-Easterly direction, resembling the valley which accompanies the dyke at Cadnant, and similar to that which is occasionally seen along the course of the dyke last described. A considerable mass of trap is found at the spot where this line is intersected by the road from Holyhead to the South Stack, and another, of precisely the same character, in the road leading from Holyhead to Port-Dafreth (635, 636.), where the schist in contact is flinty (637.). " The pyroxene" is here " completely characterized, the cleavage evident, and the crystals may be extracted. This is a most beautiful specimen of dolerite, the same as that at Mount Meissner." *Prof. Cordier.* I met with similar trap in three other localities, along the line of bearing, but the intervals are interrupted by schist.

The two dykes which lie to the N.E. of this, in Port-Newry, are composed of a coarser and tougher basalt, resembling the hard portions of the dyke at Port-Dafreth. The largest is eighteen feet wide (638.), the smallest only one and a half (639.), and they run parallel to each other, from W. of N. to E. of S., with thirty-two feet of schist interposed.

O. 5. The dykes on the Western side of Dulas habour are composed of small grained white felspar, and black pyroxene, and through them are dispersed patches (not distinct crystals) of brown felspar (572—574.). Some portions are very earthy (575.).

The black slate, in contact, in some places passes to a light grey homogeneous clay-stone (576, 577.), at others it undergoes no alteration. In one instance there are several detached fragments, which lie embedded near the side of the dyke. Pl. XVII. Fig. 4.

There are two dykes at the Southern extremity of the C. 5. promontory at Llandwyn, the basis of which appears to be formed of earthy chloritic matter, not to be distinguished from some of the earthy varieties of the chlorite schist (682.). Through it are dispersed crystals of liver-coloured felspar an inch in length (683, 684.). These dykes intermix with the surrounding schist, winding irregularly among it. The most Southerly of the two is soon lost in the sea, both to the East and West; but the other may be traced through several of the indentations formed along the coast, and is exhibited eight times in the cliff.

It appears unnecessary to enter into a separate detail of any more of these dykes, and I shall now merely subjoin the opinions of Professor Cordier concerning the composition of some others which he examined, and select a few circumstances which may seem the most interesting towards establishing their history (640—681.).

The character which many possess, is that of an irregular vein (681.), from a few inches to some feet in width, winding among the schist, and frequently ramifying in its course. They sometimes unite firmly with the surrounding rock, but in general, the line of separation is perfectly distinct, and a blow will readily divide them.

I detected from twenty to thirty between Beaumaris and C. 5. Garth-ferry, a distance of about two miles. It is very easy to overlook them when the cliff is low, or when they are only exhibited on the horizontal section formed by the beach. Where lichen and algæ coat over every thing, the only distinction is in their smoother surface, and more angular fracture: but this will frequently escape observation where a moment's inattention may carry us across one at a single step.

It will be seen in Pl. XVIII. Fig. 4. at *a* and *b*, that there is a deceptive appearance as though these dykes terminated abruptly downwards; but in these cases the course may be considered as tending upwards, obliquely to the plane of the paper, when placed vertically, and coming from some point behind it. The deception arises from the face of the cliff intersecting the inclined side which bounds the furthest extent of the dyke to the East, a fact which I verified in one instance, by removing the schist from below. There is a marked distinction in this apparent mode of termination, and that which is seen in Pl. XVIII. Fig. 2. at *a*. In the latter case, the fissure containing the basalt, gradually becomes thinner towards the end, in the former, the entire width is preserved.

The specimens examined by Professor Cordier, from this neighbourhood, were "dolerite. The pyroxene very evident, with *fer titané.*"

"A basaltic lava, but more felspathic than the others. The felspar has the filamentous character of volcanic products, resulting from the crystals being flattened. To see this, two sides of the specimen must be placed at right angles to each other."

The appearance of the flattened crystals is common to several of the very compact dykes, and may be seen in some parts of the one near Cadnant, towards its Western termination. In the small dyke Pl. XVIII. Fig. 2. *a*, these crystals are few, and extremely minute (642.), the basalt being more remarkably fine grained and tough, than any other which I met with in Anglesea.

An evident intermixture often takes place, between the trap and the surrounding schist, along the line of junction, which sometimes resembles the gradual blending of two different colours in a mass of striped jasper (642.). Small portions of schist are embedded, near the sides of the dyke, which intermix with the trap, and modify its appearance and composition (641, 643.) The schist, in contact, has frequently a blistered aspect, with irregular cavities and flaws (644, 645.).

Dykes immediately to the North of the Menai bridge.

"Dolerite with *superbe* pyroxene." *Prof. Cordier.*

"Felspar and pyroxene with crystals of pyrites (665.). The circumstance of having crystals of pyrites, though rare in streams of basalt, is easily accounted for in a dyke. The extended surface presented to the air by the stream, would enable the sulphur to evaporate, but in the dyke it is condensed. Perhaps also the dyke never came to day." *Prof. Cordier.*

The presence of pyrites, frequently in the form of distinct crystals, is common to *most* of the dykes in Anglesea.

Dykes to the South of the Menai bridge.

"Basalt very rich in felspar." *Prof. Cordier.*

"Basalt poor in pyroxene." *Prof. Cordier.*

On the South-western coast near Aberfraw.

"*Plus travaillée* than the other dykes—blistered." *Prof. Cordier.*

The passage to the earthy traps is perfectly insensible, and portions of the most genuine basaltic dykes are frequently of this nature.

At Llangwyfan—"*Wacké endurcie.* It is full of green earth, c.3. and ought to become cellular in an acid." *Prof Cordier.*

Most, if not all, of the varieties of trap included in the dykes of Anglesea are occasionally amygdaloidal and porphuritic. Some contain nodules of crystallized carbonate of lime, which do not always exhibit the usual appearance of a rhomboidal cleavage common to the whole nodule, but possess an uneven fracture, although the specimen is perfectly pellucid, approaching the character of saccharine marble (658, 659.). Embedded crystals of felspar are more common in the compact and earthy traps (661.), than in the crystalline (660.).

The compact portions, of several dykes, assume a confused appearance of crystallization, and break into small fragments, a few inches in diameter, bounded by perfectly smooth surfaces. Several of these form accurate rhomboids (676.), others exhibit this figure modified by a diagonal cleavage (675, 677); but it generally happens that their figure is less regular, and that no two faces are parallel to each other (678, 679.)   The effects of decomposition frequently extend to a considerable depth in the dyke, and we find each of these fragments, partially decomposed, presenting a portion of unaltered trap in the interior (680).

Granitic Districts
(*including the Granite and Greenstone*).
{Nos. 789 to 844.}

A rock formed of quartz felspar and mica, is found in each of the tracts laid down as including the granite; but the mineral character of the whole district is far from uniform.

Gr.   In the Southern portion, about the neighbourhood of Gwalch-mai and thence towards Llanerchymedd, the surface is broken by small detached rugged eminences, rising through a marshy ground, which is bounded East and West by an abrupt termination of the stratified rocks.

The external character of all these protruding masses is so very similar, that it is impossible to calculate beforehand on what may be the real composition of any one in particular. On examination they are found to vary extremely; one may be a true granite, the next a pure quartz, the third a greenstone, &c.

A better notion of this variety of composition may be obtained by referring to some of the specimens which were procured in the neighbourhood of Gwindu, within four miles North and South of the Inn.

Among the granitic rocks the quartz is generally white; the felspar is either white (724, 725.), brownish yellow (726, 727.), or flesh-red (728—737); the mica silvery white (725.), black, or green (730—732.). In the latter case it becomes associated with chlorite, which in many places entirely supersedes it (739, 743.), tinging both the quartz and felspar of a greenish hue (740.). The chlorite also mixes with hornblende (741—744.), and these two substances frequently predominate so much as nearly to obliterate the quartz and felspar (753, 754.). Sometimes the felspar, of a flesh red colour, forms the basis of the rock, and the other ingredients are sparingly dispersed through it (728, 729.), (745, 746.). In other places, chlorite and mica supersede the rest (752.), and we then find only patches and veins of felspar and quartz, completely enveloped in the more trap-like rock (750, 751.). A beautiful variety is composed of dark green hornblende crystallized in large plates, and intermixed with irregular patches of white felspar (755.), which however frequently assumes a greenish tinge (756.). At the same spot there are patches of crystallized carbonate of lime penetrated by yellowish green spiculæ of epidote (757.), a substance pretty generally diffused through the surrounding rock, either in veins (758.), or interlaminated with the hornblende (759, 760.). It occurs also in compact masses, intermixed with quartz (761.). Patches of genuine basalt are scattered throughout the district, completely enveloped by the granite, and possessing the same character as the trap found in the dykes of various other parts of the Island (762, 763.).

All these varieties are highly crystalline; but with them we find rocks of another description, whose composition is more nearly homogeneous. They possess a flinty aspect approaching to hornstone, and are of various shades of white (766, 767.), grey (768.), or green (769.). Here and there a crystalline structure is exhibited, or a few crystalline specks lie dispersed through the compact base.

This variety in the mineral composition is chiefly confined to those parts of the district which present a broken rugged outline. In the elevated ridge which stretches from Gwalchmai to Lanfaelog, the character is more uniformly granitic and the surface of the ground unbroken. The quartz and red felspar have not the distinctly granular appearance which they generally assume in substances of this nature, but are intermixed with a more pasty aspect than usual (734.), and the lustre frequently deadened by a superabundance of the oxide of iron (737.).

Gr. 4. The Northern district occupied by the granite is not so variable in its character; the usual appearance being that of an irregular and large grained intermixture of quartz, white felspar and silvery mica. A greasy lustre is frequently given to portions of this granite, which apparently arises from its being contaminated by a considerable quantity of talc (790—801.).

By referring to the Map it will be seen that there are two districts which consist entirely of greenstone. The general character of the rocks which compose them, is so nearly allied to some parts of the granitic district to the South of Gwindu, and their relation to the surrounding strata so very similar, that little doubt can exist of their belonging to the same formation.

Gr. 3. The district to the North of Llanerchymedd is marked by rugged, and rudely shaped masses, projecting through the surface. These extend from a spot about one mile to the North of the town, on the West of the road to Amlwch, towards the North of East, and pass a little to the North of Llandyfrydog. A prevailing character is that of an hornblende rock, composed of large crystalline plates interlacing in various directions, and cemented together by a little felspar and carbonate of lime (689.). The felspathic cement gradually increases (690.), till it forms a greenish compact basis, through which the crystals of hornblende are dispersed (691.). Other varieties present a more perfect intermixture of these two ingredients (692 — 694.), with the addition of small shining plates, apparently diallage (695.). White felspar and dark grey hornblende form also a finely granular compound, which resembles some varieties of the trap included in the dykes (696, 697.). Distinct, green crystals of hornblende are embedded in a basis of crystallized white felspar, and it is worthy of remark, that some of these crystals have been broken, and the fragments lie in different directions, surrounded by the felspar, the edges of their corresponding extremities tallying with each other (699.).

The greenstone to the East of Llanbabo is not well exhibited. Its characters are precisely the same as the former (712, 713.).

Having proceeded thus far with the description of the granitic districts, before any attempt is made to establish the probable history of the rocks which they include, it may be here remarked, that the phenomena which accompany them are so very similar to those presented by the trap dykes, that we can have little hesitation in ascribing their origin to the action of the same cause. This circumstance is premised, that the object may at once be seen for which any particular appearance, tending to establish the theory of their common origin, is recorded. From what has been stated, it will readily be conjectured, that the theory alluded to is that which ascribes the formation of these rocks to the influence of volcanic action, and it must be perfectly unnecessary to recapitulate the arguments which have been urged by others in its support, and drawn from appearances similar to those described under the details of the several dykes enumerated in the preceding part of this paper. They are such as will suggest themselves to every one, and some speak so strongly in its favour, that it seems scarcely possible for the most sceptical on this head not to allow the force of their evidence.

In addition to those arguments which may be deduced from such phenomena, it may be stated, that the number of these dykes must be very considerable; for many of those enumerated have become exposed, by mere accident, in the different quarries opened for the purpose of repairing the roads, and it may reasonably be expected that there are very many others concealed beneath the cultivated surface, as well as several which have escaped observation. In no one instance does it appear, that they are in any way associated with a superincumbent mass of the same nature, and indeed the great variety of mineral character which they assume is alone a strong argument against supposing them ever to have formed members of a common body None of the veins and fissures which contain them appear to terminate downwards, whilst on the contrary it should seem, that, in some instances, their termination upwards has been clearly ascertained. In others, there is every probability, that a considerable part of their course lies beneath the rock with which they are in contact at the surface.

There is one argument brought against the igneous theory which may be supposed to derive weight from the investigation of Anglesea. This is, that the trap, if projected in an ignited state, would have produced results of a more uniform character, whereas in many cases it should even seem that it has produced no alteration whatever upon the surrounding rock. Now, one decided example of alteration should speak more plainly towards establishing the nature of these dykes than any negative argument which might be drawn from those cases in which no such alteration is found to take place; for we know it has been determined by experiment that certain rocks, when fused, will afterwards return to their former state, if placed under those very circumstances which most probably must have existed at the time of their fusion.

This fact may be illustrated by referring to the phenomena which accompany the dyke at **Plas-Newydd.** The alteration which there takes place in the surrounding rock, although of the most decided nature, is by no means uniform, even in the same stratum. On one side, we have a mass of soft clay shale assuming a hard jaspideous character, whilst, on the opposite side, this alteration is partial, and the rest puts on a crystalline structure; and intermixed with this we also find some portions in an earthy state. That the whole is not crystallized may readily be accounted for, by supposing a superabundance of calcareous and argillaceous particles, above the requisite proportion necessary for the formation of the crystals; still, however, it shews us that in the very spot where the change is the most marked, it is yet possible for some portion to remain unaltered.

If the granite of Anglesea be justly ascribed to the same class of rocks as those which compose the trap dykes, it should seem equally certain, that some portions of it must either have resulted from the fusion of the surrounding strata, or else have been considerably modified by an intermixture with them, and consequently that it is more recent than any with which it is associated.

At the South-western termination of the Northern granitic district, there is a patch of old red sandstone. Although the ·o.·4. whole of this appears to have been considerably changed from its original character, and to have assumed a more compact and crystalline structure, yet at the furthest point West from the granite, it is evidently composed of sandstone mixed with coarse breccia containing pebbles of quartz and slate (784.) The strata run directly towards the granite, and several opportunities occur of examining the alterations which they sustain. Upon approaching the granite, the crystalline character increases, the materials become more firmly cemented, and pass into each other (785—788.), till at length, without any abrupt transition (789.), the strata merge into a crystalline rock (790—796.), in which the nodular concretions of quartz have scarcely lost the aspect of pebbles (790.). Felspar, frequently of a talcose or steatitic aspect (792.), forms the principal ingredient of the resulting granite (798—801.), which contains large, distinct concretions of quartz and mica. Through it there are also dispersed irregular masses of impure adularia, which cleave with great facility (797.). At the spot where the sandstone has first assumed the decided character of a granite, there occur a few specks of galena (796.).

A repetition of similar appearances may be traced along the boundary between the old red sandstone and South granitic district. Towards its Northern termination, immediately to the West of Llanerchymedd, the granite is found in the bottom of a valley which passes between two portions of the sandstone At the last spot where its effects are distinctly marked, there is a quarry in which the rock may truly be said to constitute the intermediate passage between the two. The remains of a coarse sandstone are evident in some parts of the quarry, and the passage (777—780.) to a perfectly crystalline mass (781.) distinctly visible. The whole shatters into small fragments, and a considerable portion is converted to yellow ochre, which also coats over the natural cleavages in the more solid portions. A vein of crystalline quartz, one inch and an half thick, traverses the decomposing, porous portion of the rock (783.), and with it are intermixed irregular stripes of chlorite, which penetrate the quartz, and are so disposed, as to form rudely parallel lines inclined to the sides of the vein at an angle of 45° (782.). Where the chlorite is not present, the quartz still preserves a tendency to cleave in this direction; a circumstance which bears a striking resemblance to a fissile texture, oblique to the disposition of a bed.

About one mile to the South of this spot, there is one of the localities already pointed out, for the fossils in the old red sandstone. The quarry in which they occur is on the confines of the granite, and the neighbouring mass of rock, which projects a few yards to the East, is in fact completely crystalline. Some parts of the quarry also approach the same structure, and a gradual obliteration of the fossils is the consequence. The impressions are coated with oxide of iron (369.), and as the matrix loses its original character, their position becomes marked only by an irregular cavity, retaining a partial impression of some portion of the cast (370.), till at length the spot where they formerly existed is simply traced by a shapeless ferruginous patch (371.). Having found the impressions of anomiæ in the midst of the altered shale at Plas-Newydd, and even where it had assumed a crystalline structure, it did not seem improbable, that some traces of these shells might also be met with in the neighbouring granite, derived from the altered sandstone. But the search proved fruitless, and indeed we can scarcely expect that any such can exist. The less compact state of the sandstone, and the character of the resulting rock, so much more uniformly crystalline than that of the altered shale, would render it less likely that any appearance of this description could be preserved. It may seem singular that I should have searched in the granite for a fossil, as a circumstance likely to increase the number of arguments in favour of its igneous origin. But what has been stated may serve as a caution against forming any hasty conclusion to the contrary, should such a discovery be ever made.

The altered appearance of the old red sandstone, which lies to the West of the Southern granitic district, was remarked in the description of that rock. The facts which have been just stated seem to point out a cause adequate to the explanation of this circumstance, and there are besides some other particulars connected with its history, which tend materially to o. I. confirm this supposition. From the lake at Llanfaelog, to Llanfihangel, the surface is swampy and uncultivated, through which many masses of bare rock project. Several of these present an aspect so highly crystalline, that at first sight a question might easily arise, whether we were not still in the midst of the granite, until a second blow from the hammer clears up the doubt, by exposing a mass of hardened sandstone. In short, the state of this sandstone appears to be only a degree removed from the more crystalline structure of the granitic district which lies to the North of Gwindu.

Near Llanfihangel church, on the South, and in the midst of an assemblage of rocks distinctly composed of the brecciated materials, we find a mass of trap (804, 805.). The felspar is sufficiently distinct, and forms the chief ingredient in the basis of the rock, through which a few embedded crystals of the same mineral are scattered, giving it a slightly porphyritic character. The whole assumes a greenish tinge, but the colouring substance does not appear to be of a very crystalline nature, and is probably chlorite. This intermixes with a confused aggregation of hornblende and diallage (807, 808.), passing by insensible shades to the breccia which surrounds it. In the very midst of the more crystalline portion we find small patches possessing a trace of the original character not quite obliterated (809).

The rock often passes also to a light green felspathic mass, spotted or mottled with dark green. Several such appearances occur in the form of smooth nodules, already alluded to as embedded pebbles, but I strongly suspect that they must be of a concretionary nature, similar to those in the steatitic rock near Bangor.

The whole of the exposed rock to the S.E. of Llanfihangel church, has more nearly the external character of a mass of trap than of any other substance. It possesses no traces of stratification, but is rent by fissures which divide it into prismatic and rhomboidal blocks. One of these is so singular in its appearance, that I have given a sketch of it, Pl. XVIII. Fig. 5. It resembles a basaltic column lying upon its side, and is composed of felspathic and chloritic matter, mottled and blended together (810, 811.).

On the N.W. of the lake near Llanfaelog, there are several instances of a similar passage of the breccia to a trap rock (812.).

This apparent conversion of the schistose breccia, belonging to the old red sandstone formation, to a trap rock, seems more distinctly to connect the greenstone with the granite, and to point out a common origin for the two, which also receives additional confirmation from the examination of the tracts occupied by the former rock. The patch of greenstone to the North of Llanerchymedd is surrounded by greywacké, the basis of which is a glossy black clay-slate. In the immediate vicinity of the greenstone, this greywacké is curiously affected; the embedded fragments of schist assume a yellow decomposing tinge, whilst the quartz becomes more crystalline (701—705.) The next step presents a rock of decomposing aspect, through which are scattered traces of crystalline structure, resulting from an imperfectly formed hornblende, mixed with felspar (706—710.). The latter is distinctly marked, but the crystals of the former bear a strong resemblance to fragments of slate. They are frequently broken transversely, a circumstance which it has been stated also occurs in the genuine crystals of the same substance in the neighbouring hornblende rock.

It does not appear very evident why hornblende should here result from the fusion of schist, and that pyroxene should be a constituent of the dykes which are presumed to be of similar origin. There is, however, one point of difference between them. In the dykes, the fused matter appears to have been injected into a fissure of the superincumbent rock, but in the present instance the alteration has taken place without any progressive motion. There are other rocks, in this part of Anglesea, of which hornblende is an ingredient, where the transition from the schist to the trap is not marked by a distinct line, and where a similar explanation might be given of their origin.

Near the summit of Llaneilian mountain, towards the South, we find masses of this rock, protruding through the greywacké, in which the hornblende is sometimes well crystallized (715.), and at others scarcely to be detected (716.).

At the bottom of the cliff, to the N.E. of the highest point of this mountain, a similar rock is found, but the hornblende is not so distinct as in the former case (717.). Upon ascending the cliff the appearance of a dyke is gradually lost, and it scarcely exhibits a structure sufficiently crystalline to separate it from the schist (719—722.). Through this dyke there run several veins of quartz, which also abound in the surrounding rock, a fact which I do not recollect witnessing in any other dyke in Anglesea. Irregular strings of reddish compact felspar, of a cotemporaneous character, are also found in it (718.). The schist in contact is a fine grained clay-slate (723.), and in the dyke there occur several strings, or thin laminæ, of clay-slate of the same nature.

Patches of glossy crystalline clay-slate are also found among the hornblende rocks to the North of Llanerchymedd.

On the South side of the road from Llanerchymedd to Llechynfarwy, before quitting the former place, there is a quarry which partly consists of clay-slate, and partly of hornblende rock and greenstone, similar to that on the North of the town.

Considering the extensive influence which must have been exerted to form the granitic districts, we might also expect to find the rocks in their vicinity modified by its action. Where the South granitic district joins the older rocks to the East, it is not so easy to ascertain when an alteration has taken place; since we are not always certain of the original character which these rocks themselves possessed. In some places, however, there appears to be little doubt of the fact.

On the sea-shore, immediately South of Llanfaelog lake, the confusion and alteration impressed upon the schistose rocks are of a very marked description. They vary in composition and aspect at every step, and have scarcely a trace of fissile texture remaining. There are slight appearances of a crystalline rock, resembling some varieties in the granite round Gwindu (824.), but the general character is that of an homogeneous flinty mass of different shades of green (825, 826), grey (827.), and brown (828, 829.). Since all these will fuse before the blowpipe, though with difficulty, they seem to approach the character of a hornstone. One of the specimens presents a singular fact, and as the experiment was several times repeated, there can remain no doubt of its accuracy (827.). It forms a dirty white mass between compact and finely granular, and seems to consist principally of quartz, but contains also a little lime disseminated through it. A few faint streaks of green matter, resembling chlorite, are intermixed with the substance of the stone. When the specimen is exposed to a red heat, the green veins immediately turn jet black, assume a laminated texture, and strongly resemble pyroxene. It then fuses, with some difficulty, to a black glass. There are some portions of a more compound nature (830.), intermixed with the homogeneous rock, which appear to be composed of small fragments of quartz firmly embedded in a paste of the same substance (831.), and others in which the embedded fragments are so loosely set, that they might be detached (832, 833.). With these are associated patches of blistered schist, gradually blending itself with the compact mass (834.). Proceeding Eastward along the shore, we find traces of a laminated texture making its appearance. The whole rock still forms a flinty mass, but the smooth surfaces exhibit parallel contorted lines of obscure yellow upon a green ground (835—837.). This character prevails until we arrive at Llangwyfan, after which the rocks become more regularly schistose.

There are numerous trap dykes scattered throughout the whole of this district (813—823.). These vary considerably in character; some form a perfect basalt highly charged with crystallized carbonate of lime, tinged green (813.), but the generality, are of a more earthy nature, and vary through different shades of dark grey and green. In texture and composition, they often resemble clay-slate so closely, that a detached specimen might easily be mistaken for this rock (820—822.). They are generally porphyritic, containing embedded crystals of felspar, and alter their character completely, and suddenly, through different parts of their course. Many of them are seen to terminate in both directions, and some form mere bumps rising through the hardened schist, and are themselves again intersected by smaller dykes of a different character.

There are several other appearances of a similar description impressed upon the schist near the granite, both in the neighbourhood of Llanerchymedd, and of the North granitic district. In one spot, about a mile and an half to the N.E. of the Paris mountain, it is intermixed with veins of a granitic description. They consist principally of dull mica, which is associated with felspar, quartz, and a steatitic substance (802.). The surrounding schist assumes a hardened aspect (803.).

Perhaps we may also ascribe the flinty beds dispersed throughout the schist, to the West and S.W. of this spot, to a change of character impressed by the granite. If so, the chert, which traverses the Paris mountain, will rank with them; and the decomposing schist, which accompanies it on either side, must be ascribed to the more partial influence of the same action. The external character of this mountain is very striking. On the N.W. it slopes gradually from the top, but to the S.E. it presents a precipitous side, from which project the edges of the schistose laminæ, to all appearance as sharp as though they had scarcely sustained the action of the weather since they were first placed in their present position, although the materials of which they consist are in a very decomposing state.

Although the circumstances detailed above seem to indicate that the granite and greenstone have been derived from the fusion of the stratified rocks in their neighbourhood, still, the marks of violence and disturbance which accompany them, tend further to shew, that some portions of these rocks have been protruded from below. The structure of Llaneilian mountain affords an interesting and important illustration of this fact, and forms a prominent feature in the Geology of Anglesea.

In Pl. XIX. Sect. B, we have the greywacké dipping from the granite on the West, and succeeded in order of superposition by the chlorite schist, dipping also in the same direction. Upon referring to Sect A, which is exhibited on the coast, the greywacké is found to be terminated abruptly to the West by a vertical fault, which explains the apparent anomaly of the superposition of the older rock. This greywacké also reposes unconformably upon some black clay-slate of the same series. The explanation which suggests itself is, that the granite has removed the greywacké from its original position, and that the lowest beds of the portion removed, repose towards their termination Northwards upon the superior beds of the same series. The hollow tract which runs across the mountain, between the summit and the highest point of granite, is composed of glossy black clay-slate, intermixed with patches of quartzose rocks, which project in irregular masses. The lowest portion of the removed greywacké is a green clay-slate, which may have formed either the lower beds of the greywacké series, or the upper of the chlorite schist. It is much shattered; a circumstance which causes the alluvial matter to collect, and consequently the line of demarcation, between it and the black slate below, is distinctly marked by the vegetation which covers it. A deep ravine runs from the Northern termination of the granite to the sea-shore, from which part of the removed portion may have been derived. The most Westerly, and therefore the uppermost beds of the disturbed greywacké, are composed of the same black clay-slate as that upon which the lower beds repose unconformably.

Another circumstance which seems to have resulted from the intrusion of the granite, occurs to the East of this spot. Descending from the Eastern summit of the mountain towards Dulas harbour, the ground forms a gradual declivity, broken by projecting hummocks. These, as well as the rocks on the shore, consist of a most heterogeneous mixture (857—867.). Hardened sandstone (863—865.); clay-slate, which shivers into sharp hard fragments, each tarnished with a glossy coat (866, 867.); large masses of quartz and schist, rudely intermixed and bound together by a basis of fragmental matter (859—862.).

Among this confusion there are patches of purely crystalline rock, consisting of red felspar, quartz, and chlorite (857, 858), which appear to assimilate this tract to some portions of the district round Gwindu. This conglomerate is seen, at its Northern G.1. termination, reposing upon the clay-slate, which assumes a compact quartzose aspect (852.), and contains concretionary nodules (851.). In one spot it consists of crystalline quartz, through which are dispersed numerous fragments of slate (853.). These fragments have, in many instances, sustained an alteration in character, and become blended with the quartz. On a weathered surface they decompose, the quartz assuming a cavernous appearance.

By referring to the sections it may be seen, that the intrusion of the granite in the two principal districts, has produced effects in opposite directions. In the Northern district, the dislocation and tilting of the schist lies on the West, the older members having been upheaved and brought to the surface on that side of the granite, whilst on the East we find the newer rocks comparatively in a state of repose. In the Southern district the reverse is the case, and the upheaving of the mica and chlorite schists has taken place to the East, the old red sandstone being in contact to the West.

Having examined the phenomena which accompany the granite and trap, the next endeavour will be to explain certain appearances where a cause seems to have been exerted, similar to that which produced these rocks, though without the actual protrusion of any volcanic product.

A reference to the localities noted on the Map will be sufficient to shew that there can scarcely be any part of Anglesea which has entirely escaped the influence of an action so general as that which formed the numerous dykes seen bursting through so many parts of its surface, and we may naturally expect that some places must exhibit traces of this influence, where the results are more equivocal than those already described.

At Carnel's Point, the appearances so closely resemble those exhibited on the confines of the granitic districts, that it seems scarcely possible to ascribe their origin to the action of a different cause. Upon approaching the Point, a little to the North, the greywacké is associated with a rock, which at the time I mistook for a conglomerate of rolled pebbles, but which is composed of concretions running together in the same manner as the steatitic rock near Bangor (870.). This passes to a perfectly crystalline mass formed of quartz, felspar, chlorite, and mica (871—878.), in which the traces of a concretionary structure are sometimes evident (871.). In other places the transition to the earthy rocks which are contiguous, is sufficiently marked (880.). The character of the surrounding schist is singular. In part it appears to have sustained no alteration, but the greater portion assumes a yellow decomposing aspect, and in some places it passes to a hard semi-jaspideous mass (883, 884.). Rather large crystals of quartz are found attached to compact masses of the same nature, which are dispersed through these rocks (881, 882.).

Another spot, which has strongly the appearance of having been subjected to some violent disturbing cause of the same nature, occurs at Moel-y-don ferry, on the Cærnarvonshire side of the Menai. The strata of limestone and grit, belonging to the magnesian limestone, are found confused, tilted, and, for several yards, disposed in a most disorderly manner. The alteration which the substances composing the strata undergo, is also of a marked description. The sandstone, which in other places is red, becomes white (452.), hardens, and passes to a compact siliceous stone, resembling white flint (453.). In some places it approaches the common dark chalk-flint (454.). It is intermixed with crystallized limestone and bitter spar (456—462.), and some portions of the specimens are in a pulverulent state. The red limestone either becomes very compact and crystalline (463—467.), passes to a brown bitter spar (469.), or assumes the character of a nearly arenaceous white limestone (470.). About the centre of the disturbed portion, the materials of the several strata are mixed together (471—473.), presenting a singular scoriaceous appearance (474.). The dyke which crosses the Menai at this spot, intersects the line of the disturbed portion at right angles. A little to the North of the ferry where the disturbance ceases, the sandstone, which is fine grained with small pebbles dispersed through it, appears to have been hardened and turned white (451.), and is here quarried as a whetstone.

The present disposition of the stratified rocks, from the greywacké upwards, seems strongly to favour the hypothesis which ascribes the formation of the granite and trap to volcanic action. If we suppose the different portions of each formation to have been once continuous, their original bearing appears to have been from N. E. to S. W., and their dip towards the S.E. In every case they terminate with great abruptness against the older rock, and near the junction they frequently appear to have sustained some violent action.

The singular transition from old red sandstone to quartz o. s rock, on the summit of Bodafon mountain (p. 391.), resting upon hornstone derived from the clay-slate (p. 373.), appears to be the Northern termination of some volcanic influence which extended Southwards to Llangefni, and from thence, towards the S. W. to Aberfraw. Hand specimens cannot convey a just idea of the appearances exhibited in a quarry of this hornstone. The irregular intermixture, which takes place in the different shades of colour, strongly resembles the result afforded by agitating the ingredients of a semi-fluid mass, and the solid horn-stone passes to a blistered schist.   Proceeding Southwards, we come upon a peculiar variety of chlorite schist, in which the quartz forms a homogeneous basis, and the chlorite or mica lies disposed in parallel laminæ (p. 369.).   At the spot where this occurs furthest to the North, it forms a mass of rock scarcely protruding above the surface of the swampy ground on the East of Bryngole, and is not easily accessible.   From hence, some patches of schist are found which pass to jasper, and others to the same translucent, green quartz rock, with a glo-bular structure as that at Llangefni, (p. 384.).   In an inter-mediate state, the schist has a·fragmental aspect, the fragments drawn out into strings (243.).   At Bodorgan, we find a mass of basalt and greenstone protruding in the midst of the small patch of schist which rises through the grit (p. 397.).   It should seem therefore, that the disturbing force, which cut off the further extent Westward of the mountain limestone and coal-measures, has acted upon the old red sandstone, clay-slate, mica-slate, and chlorite-slate, and that the respective results are quartz rock, hornstone, homogeneous quartz with scales of mica or chlorite, and jasper or translucent quartz rock.   To this list may be added another substance, equally singular.   To the West of the spot where the chlorite schist passes to quartz rock, North of Llangefni, the grit rises abruptly to the East of the river, and through it there protrudes a rude tabular mass of white quartz rock (868.), which may be supposed to have re-sulted from an alteration of the grit, (see Pl. XX. Sect. H).

C. 3.   The district lying North and South of a line from Gwalch-mai to Llangefni, will, by the above supposition, be included between two distinct modifications of volcanic action, which were probably united beneath it.   The result, as might be expected, is the utter confusion and complete alteration of the intervening rocks, the appearance of which has been already described (p. 372.). Among other phenomena which tend to confirm this hypothesis, is the occurrence of several large hummocks of white quartz, scattered over the surface in the neighbourhood of Trefdraeth (869.). They consist of sandstone passing to quartz rock, and are possibly the remains of grit, from the strata belonging to the coal-measures; and if so, it is equally possible, that the hard compact limestone found with the jasper may have been derived from the same source.

A similar explanation will apply to the other districts of this nature, and particularly to the jaspideous ridge which extends from Llanfinnan to Red-wharf bay, (p. 372).

There are a few circumstances connected with the history of the serpentine which render it probable that this rock is the result of an action impressed upon some of the limestone beds, dispersed through the chloritic districts.

The general outline of each of the serpentine districts, presents an aspect of great disturbance, want of stratification, and other appearances usual in a trap formation. At the Eastern termination of the Southern district, the transition from the chlorite schist to the old red sandstone is abrupt, and on the Eastern side of the line of junction we meet with a trap rock, apparently derived from an alteration impressed upon a portion of the latter formation, a description of which has been given, p. 432. The connection between serpentine and greenstone is equally remarkable at Llanfechell. The patches of serpentine run S. W. from the principal quarry, and with them are found some which seem intermediate between that rock and greenstone (589—591.). Others pass from a compact and crystalline variety of the latter, including veins of serpentine and epidote, to a more earthy rock, with a greasy chloritic aspect, containing but few crystals of pyroxene. The chlorite schist in the neighbourhood is much hardened, especially in the high ground to the S.E. of this spot. Veins of epidote are found where the rock approaches the character of greenstone (592—596.).

C. 3.   To the West of the pool at Gwalchmai, the chlorite schist is intermixed with limestone, and a dyke intersects the rock close to a small bed of this mineral. The natural fissures of the dyke are coated with an unctuous substance resembling serpentine (584.), and some portions of the dyke itself have the earthy chloritic aspect of the greenstone in the serpentine at Llanfechell (585.). Veins of crystallized carbonate of lime are found in it, and when the primitive rhomb is extracted from them, two of its faces appear finely striated parallel to the longer axis (586—588.).

Conglomerates.

{Nos. 885—913.}

There are two tracts laid down in the Map under this denomination, one of which has been already alluded to, as its connection with the North granitic district appeared to afford an explanation of its origin. The other, which occurs at the most Northerly point of Anglesea, possesses a few characters in common with the former, but there is no evidence to shew that it has resulted from an action of a similar nature. It consists of a most confused intermixture of chlorite schist, greywacké, clay-slate, large masses of quartz, and limestone, with several traces of trap, Pl. XVIII. Fig. 6.

To the East of Cemmes, the cliff consists of yellow ochre and steatite (895.), both of which are there quarried as articles of commerce. They evidently result from the decomposition of the schistose rocks, and the gradual passage from the solid to the earthy state is distinct (892—894.).

The large masses of quartz, which are dispersed through the

district, have sometimes an homogeneous aspect (889.), but in general they retain the traces of a coarse breccia (885—887.), or sandstone (888.), from which they seem to have been derived. In the midst of the steatite there are some which possess a porous earthy aspect (896.).

The limestone, which intermixes with this conglomerate, is the same as that about Llanfacthlu, and belongs apparently to the chlorite schist (906—913.). It does not appear to have sustained any alteration, except in one or two places where it is in contact with some trap (908.). In one spot to the West of Cemmes it consists of small irregular black globules firmly cemented in a basis of rather lighter colour, which at first sight resembles the traces of an organic structure (909, 910.). In another spot where it possesses the same shaly character as at Llanfacthlu (912.) there are slight appearances resembling anthracite (913.).

Although the ingredients which compose this conglomerate cannot be said to lie in distinct strata, still there are several places where a rude kind of alternation is visible, somewhat like the arrangement which often takes place in a mass of diluvial matter These rudely formed beds indicate a considerable dip towards the North; from which it should seem, that the conglomerate succeeded the chlorite schist in order of superposition, and that the present inclined position was impressed upon each at the same time.

Diluvium.

{Nos. 914—980.}

Deposits of diluvial matter are scattered over each of the chloritic districts which bound Anglesea to the West. They occur in the form of very obtuse conical hills, the diameters of whose bases varies from about a quarter to half of a mile, and the surface, in consequence, presents a succession of gentle undulations clothed by cultivation, the subjacent rock being exposed only in the hollows between them. The internal structure of these hills is exhibited in several places along the coast to the South of Llanfacthlu, and also between Monachdy and Wilfa. They consist of fine grained sandy materials deposited in layers, and in several instances it happens, that the coarser ingredients form the uppermost portion. The rolled pebbles are not numerous, and there are a few large blocks of schist, greenstone and limestone, dispersed through them.

C. 5.   From Penmon to Beaumaris there is a low cliff of diluvium, through which are dispersed numerous large blocks of limestone and grit, derived from the coal-measures. This terminates immediately to the South of Beaumaris, in a mass about sixty feet thick. The character of this diluvium does not resemble that which is found to the West of the Island (where it bespeaks a succession of deposition), but forms a rude mass in which the materials appear to have been brought together by a single effort. A suite of specimens has been selected from the embedded pebbles. The smallest and most rolled consist of various granites, traps, and the older stratified rocks. Some of the blocks of limestone and of the more recent strata, are of very large dimensions, especially as we approach the mountain lime at Penmon.

Alluvium.

Wherever the coast to the S.W. is low, the neighbouring country has suffered from drifts of sand, which in some places have covered up the soil to a considerable distance inland, presenting a dreary outline, broken only by a few projecting points of schist. This sand is still active in making its annual encroachments, if we may judge from the half buried walls which have been recently built over the low ground on each side of a road from Newborough towards the N.W.

In the preceding pages it has been my endeavour to relate those facts which appeared most likely to facilitate the future investigation of the Geology of Anglesea. It cannot be expected that a first account of any complicated country, should be accurate in all its details; for, in such cases, the time necessarily consumed in obtaining a clue to the examination, will seldom leave sufficient opportunity of accurately verifying all the points of relation which may exist between contiguous formations.

EXPLANATION OF THE PLATES.

———

PLATE XV

CONTORTIONS in the strata of the quartz rock at Holyhead mountain. Sketched from the South Stack .... p. 363.

PLATE XVI.

Fig. 1. Cleavages exhibited by the strata of the quartz rock ... p. 364.

Fig. 2. Vertical section of a mass of breccia (*a*), and a quartzose vein (*b*) connected with it, which rises through the chlorite schist, near its junction with the quartz rock .... p. 366.

Fig. 3. Junction of the quartz rock (*a*), and chlorite schist (*b*) to the West of Rhoscolyn .... p. 366.

Fig. 4. Section of the stratified chlorite schist .... p. 371.

Fig. 5. Serpentine (*a*) rising abruptly through the chlorite schist (*b*), which dips in various directions .... p. 376.

Fig. 6. Massive serpentine (*a*) gradually assuming a schistose character (*b*) .... p. 376.

Fig. 7. Appearance presented by the greywacké slate on the shore near Monachdy .... p. 383.

      (*a*) Hard, green, and unlaminated portion, passing gradually on one side to a schistose black slate (*b*), and terminated abruptly against a similar rock on the other.

Fig. 8. Arrangement of particles in the stratified grit at Bodorgan, p. 395.

PLATE XVII.

Fig. 1. Junction of mountain limestone and schist to the South of Llangefni .... p. 396.

(*a*) Hard, quartzose schist, scarcely laminated, rising abruptly through the limestone on the East.

(*b*) Fault, thirty yards wide.

(*c*) Rubble, consisting of fragments of the limestone.

(*d*) Soft, shattery green and red clay slate.

Fig. 2. Contorted limestone and shale, at Caint .... p. 397.

Fig. 3. Appearance of Great-Ormes Head from Beaumaris...... p. 398.

N.B. In the following figures, the shaded portions represent trap.

Fig. 4. Dyke in Dulas harbour containing embedded fragments of clay slate .... p. 421.

Figs. 5 and 6. Two sections, at right angles to each other, of a mass of chlorite schist, between Port-Dafreth and Borth-Anna, exhibiting the termination upwards of a basaltic dyke...................... p. 416.

Fig. 7. Section, ninety feet in length, along the course of the dyke which runs through Holyhead Island. The trap is seen, surmounted by the schist .... p. 416.

PLATE XVIII.

Fig. 1. Section along the course of the dyke at Cadnant ...... p. 413.

(*a*) Road from Beaumaris to the post road.

(*b*) Marshy ground to the East of Castellos.

Figs. 2, 3, 4. Sections of some dykes exhibited in the cliff between Beaumaris and Garth-ferry .... p. 422.

Fig. 5. Portion of the old red sandstone, passing to a variety of greenstone, with a columnar structure .... p. 433.

Fig. 6. Conglomerate at Wilfa .... p. 444.

(*a*) Quartz.

(*b*) Basaltic dyke.

PLATES XIX and XX.

(A to M). A series of parallel sections across Anglesea from N.W. to S.E., referred to on the Map by corresponding letters.

(N to O). Two parallel sections from S.W. to N.E., at the North-Western corner of the Island.

(P). Section at the N.E. corner.

A few changes in the mineral character of some rocks included in the same formation are referred to in the following manner:

- (*a*) Green clay slate,
- (*b*) Black clay slate,
- (*c*) True greywacké slate,
- (*d*) Micaceous schist,
- (*e*) Chlorite schist,
- (*f*) Conglomerate of jasper, limestone, &c. in the chlorite schist.

PLATE XXI.

Geological Map of Anglesea.

The principal districts included in each formation are artificially divided, for the purpose of easy reference, in the following manner,— (the order of arrangement always proceeding from West to East,) see page 360.

Quartz Rock. (Q).

Q. 1. Most Westerly portion of Anglesea, including nearly half the Northern division of Holyhead Island.

Q. 2. Also in Holyhead Island. To the S.W. of its Southern division.

Chlorite Schist. (C).

C. 1. Includes the greater part of Holyhead Island, and is bounded to the North by a line from Llanrhyddlad to Llanbabo, and on the S.E. by continuing this line through Llanfihangel to the sea.

C. 2. The Northern part of Anglesea—between Carnel's Point and Llaneilian Point.

I. S. Hawkins del.ᵗ G. Scharf Lithog.

Quartz Rock opposite the S.ʳᵈ Stack. Holyhead Island.

Printed by Rowney & Forster.

Fig. 1.

Fig. 2.

Fig. 3.

Fig. 4.

Fig. 5.

Fig. 6.

Fig. 7.

Fig. 8.

J. S. Henslow del. G. Scharf Lithog.

Printed by Rowney & Forster.

Fig. 1.

Fig. 2.

Fig. 3.

Fig. 4.

Fig. 6.

Fig. 5.

Fig. 7.

I. S. Henslow del. C. Scharf Lithog.

Printed by Rowney & Forster. London.

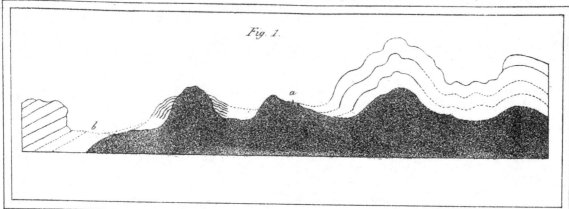

Fig. 1.

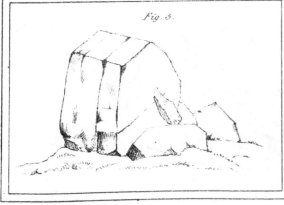

Fig. 3.

Fig. 2.

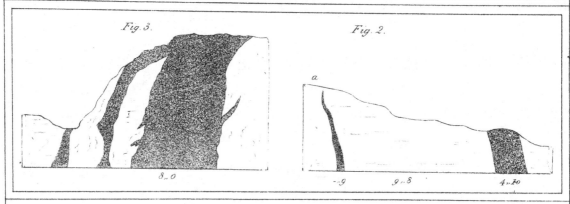

Fig. 4.

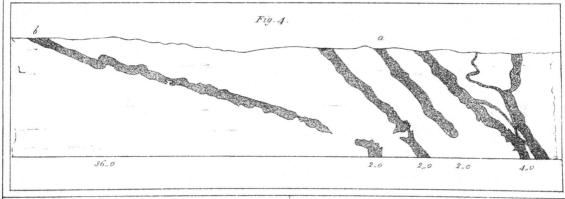

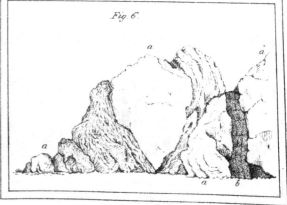

Fig. 5.

Fig. 6.

I. S. Henslow del. G. Scharf Lithog.

Printed by Rowney & Forster. London.

The material originally positioned here is too large for reproduction in this reissue. A PDF can be downloaded from the web address given on page iv of this book, by clicking on 'Resources Available'.

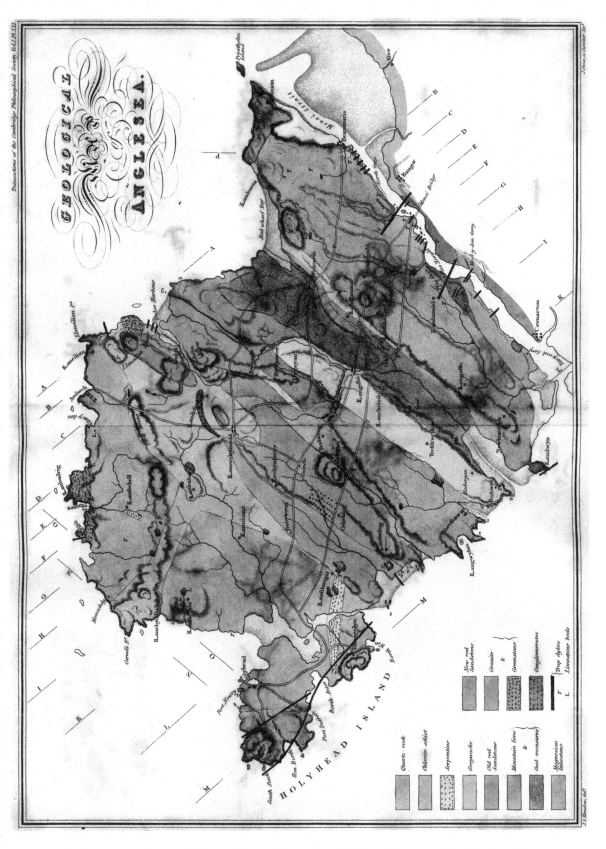

The material originally positioned here is too large for reproduction in this reissue. A PDF can be downloaded from the web address given on page iv of this book, by clicking on 'Resources Available'.

C. 3. Middle district—from Dulas on the North to Aberfraw on the South.

C. 4. Small strip to the East of the last—between Red-wharf bay and Caint.

C. 5. Western district—from Llandonna to Llandwyn.

Serpentine. (S).

S. 1. In Holyhead Island—from Rhoscolyn to Llanfihangel.

S. 2. Near the centre of C. 2.—to the S.W. of Llanfechell.

Greywacké. (G).

G. 1. Between C. 1. and C. 2., and to the East of each as far as C. 3.

G. 2. Towards the North of C. 3.—a small strip running from Bryn-gole to the S.W.

G. 3. Bounds C. 3. on the S.E.—from the North of Llangefni, to the South of Aberfraw.

G. 4. Between C. 4. and C. 5.—from Red-wharf bay to Llanfihangel East.

G. 5. Small patch N.E. of the last—at Llandonna.

G. 6. At the Eastern termination of C. 5.—West of Beaumaris.

G. 7. To the South of the last.—At Garth-ferry, and on the opposite coast of Cærnarvonshire, from Bangor to Aber.

Old Red Sandstone. (O).

O. 1. Separates C. 1. and C. 3. at their Southern termination, and runs as far North as Llanerchymedd.

O. 2. A small patch at the N.E. termination of the last—immediately South of Llanerchymedd.

O. 3. A similar patch to the N.E. of Llanerchymedd.

O. 4. Another, to the N.E. of the last—in contact with the Northern granitic district.

O. 5.   On the N.E. of C. 3.—From Dulas harbour to Bryngole.

O. 6.   A small spot, surrounded by the mountain lime, to the East of the last.

Mountain limestone and Coal-measures. (M).

M. 1.   The most extensive district of this formation—from Dulas harbour to Bodorgan.

M. 2.   S.E. of C. 5.—From Garth-ferry to Plas-Coch.

M. 3.   N.E. termination of Anglesea—including Priestholme island.

Magnesian Limestone and New Red Sandstone.

These lie to the South of M. 2.

Granitic Districts. (Gr).

Gr. 1.   Largest district — towards the centre of the Island — round Gwindu.

Gr. 2.   Small patch in G. 1., to the South of C. 2.—East of Llanbabo.

Gr. 3.   S.E. of the last, and North of Llanerchymedd.

Gr. 4.   Northern district—from Llaneilian mountain to Dulas.

Conglomerates.

One of these occurs at the most Northerly point of Anglesea—the other to the East of Gr. 4.

Trap Dykes.

These are referred to the formation in which they occur.

N.B. By an error of the Engraver, the references to these in the Map are made with a (T), instead of with a (τ) as mentioned in p. 401.

St. John's Coll.   J. S. HENSLOW.
   Nov. 26, 1821.

THE

MAGAZINE

OF

NATURAL HISTORY,

AND

JOURNAL

OF

ZOOLOGY, BOTANY, MINERALOGY, GEOLOGY,
AND METEOROLOGY.

CONDUCTED

By J. C. LOUDON, F.L.S. G.S. &c.

MEMBER OF THE ZOOLOGICAL SOCIETY OF LONDON, AND OF VARIOUS
NATURAL HISTORY SOCIETIES ON THE CONTINENT.

VOL. I.

LONDON:
PRINTED FOR
LONGMAN, REES, ORME, BROWN, AND GREEN,
PATERNOSTER-ROW.
1829.

mencement of intermittent fevers. The flowers of the G. virgàta (*g*) pessess a delicious fragrance.

Assembling, as we do, in this country, and easily procuring, every thing that is useful and ornamental, it may be a mere matter of amusement thus to sketch the uses of foreign plants : but could we, for a moment, imagine ourselves out of the reach of European assistance and European luxuries, we should be glad of every resource presented to us by nature, and congratulate ourselves on possessing a knowledge of her treasures. In all primitive countries where vegetation is luxuriant, it supplies every department of life with a mine of wealth. Not only is it employed in the construction of edifices, weapons of offence and defence, household utensils, and musical instruments; but it affords clothing, ornaments, food, medicines, and is, even, the great source from which rude people draw their poetry and romance.

The physiological botanist, therefore, possesses great advantages when travelling in these wild countries; and it has been observed, that the respect felt by savages for this sort of knowledge increases their confidence in strangers, and gives them a favourable impression of the omniscience of civilised Europeans.

ART. IX. *On the Leaves of Maláxis paludòsa.* By the Reverend JOHN STEVENS HENSLOW, Professor of Botany in the University of Cambridge.

Sir,

IN the fourth volume of the *English Flora*, Sir James Smith has described the leaves of Maláxis paludòsa, as " roughish about the extremity, often somewhat fringed, so that this plant may perhaps have given rise to the report of a hairy-leaved Orchis," &c.

This plant occurs in great plenty in the bogs on Gamlingay Heath, Cambridgeshire, where I had an opportunity of examining it a few days ago, and ascertained the cause of the fringed appearance of the leaf, alluded to by Smith. Every specimen I gathered exhibited this in a greater or less degree, and it required only the assistance of a common lens to show me that it was occasioned by numerous little bulbous germs, sprouting from the edge, and towards the apex of the leaf, as represented in the accompanying sketch (*fig.* 197. *a b*). They were of the same colour as the leaves, green on those which were more exposed to the light, and quite white on those which were lowest on the stem, and half buried in peat and moss.

Some of these germs were so far advanced as to have put forth the rudiments of two or three leaves (*c d*); others less so (*e f*).

197

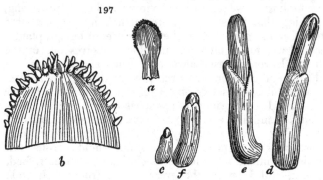

These plants often occur in little clusters of half a dozen or more close together, which may be accounted for by several of the germs arriving at perfection, whilst the rest perish. Otherwise, so far as I have observed, the plant is generally continued by a single offset, and three or four of the old decayed bulbs may be found, one below the other, among the peat, and still attached to the living stem.

This and Maláxis Loesèl*ii* are probably truly parasitic.

I remain, Sir, &c.

Cambridge, Aug. 14. 1828.   J. S. Henslow.

Art. X. *Progress of Geology.* By R. C. T.

In a former article (p. 249.), devoted to the consideration of two interesting geological memoirs, it was judged a fitting opportunity to contribute a series of preparatory illustrations of stratification, partly original, and the remainder selected from authentic data.* If we deviated somewhat from the strict course of an elementary treatise, and brought forward matter which should have formed a section at a more advanced stage, it will, perhaps, be conceded that those details were not unseasonably introduced. Our progress will henceforth be more regular. In the limited space which can here be appro-

* During the progress of this article, all the illustrations which are not expressly accounted for, may be considered as original.

To those who are familiar with the gigantic scale on which the geology of some other countries is exhibited, the examples we have selected may appear trivial; but it must be remembered that our main design is the illustration of *English* geology.

THE

MAGAZINE OF NATURAL HISTORY,

AND

JOURNAL

OF

ZOOLOGY, BOTANY, MINERALOGY, GEOLOGY, AND METEOROLOGY.

VOL. III.

1830.

CONDUCTED

By J. C. LOUDON, F.L., G., & Z.S.

MEMBER OF VARIOUS NATURAL HISTORY SOCIETIES ON THE CONTINENT.

LONDON:

PRINTED FOR LONGMAN, REES, ORME, BROWN, AND GREEN,

PATERNOSTER-ROW.

1830.

ART. V. *On the Specific Identity of the Primrose, Oxlip, Cows-lip, and Polyanthus.* By the Rev. JOHN STEVENS HENSLOW, Professor of Botany in the University of Cambridge.

Sir,

OUR knowledge of vegetable physiology has not been hitherto sufficiently advanced, to furnish us with any precise rule for distinguishing the exact limits between which any given species of plant may vary. Hence the most accurate observers often differ in their opinions, whether two or more individuals should be considered as mere varieties of the same, or be raised to the rank of separate species. Indeed, the more accurate our powers of discrimination become, the more inclined we seem to be to multiply species. There are, how-ever, certain stubborn, well-authenticated facts, which tend to lower the authority of that discriminating tact which the ac-complished naturalist possesses in so great perfection, and to persuade us that it is not unlikely that this will, after all, prove to be of little or no value towards the final solution of the question. It should now seem that nothing but the multiplied results of direct and accurate experiment can be allowed to form the basis of our speculations in this, any more than in every other department, of science. One well authenticated fact will easily shake the most specious conjectures; and, if we do not listen to Nature when she is taking her own method of opening to us her mysteries, we may be assured that we have that spirit within which would rather be led by its own fancies than bow even to the still higher authority by which Nature is herself directed. Cultivation, we know, produces numerous and very strange varieties of the same species; and, what seems highly remarkable, these varieties, when once esta-blished, frequently continue permanent through a succession of crops raised from them by seed, except now and then, where an accidental return to what is considered to be the original stock takes place. Thus, to pass over the thousand well known cases among our culinary plants, we find many seedlings of the variegated sycamore striped like the parent plant, though some of the same crop have lost this character. But if it is no sure test, that a given form should cause an in-dividual to rank as a distinct species, merely because we find that form can be propagated unaltered by seed; still less is it any clue to the accurate discrimination of species, that the same character should have been retained by the same individual for many years. On the other hand, however, it is clear that one single instance of change from one form to another, whe-ther produced by seed or culture of the individual, if well established, is quite sufficient to reduce any two of the most permanent forms to the subordinate station of mutual varieties.

What has been hitherto recorded of the production of hybrids, has rather thrown confusion than order in the way of this enquiry. The parents of many of these supposed hybrids may, after all, turn out to be no more than mere varieties of the same, though, in some instances, they have, in all probability, belonged to different species. What the law of nature in this respect really is, we can hardly be said to have ascertained. If we search for analogy in the neighbouring study of entomology, we shall find that it is only very lately that entomologists have ascertained that the nineteen varieties of Coccinélla variábilis, mentioned in Stephens's *Catalogue* as found in Great Britain, form only a single species. Mr. Stephens has recorded a similar reduction in the nineteen British varieties of Coccinélla mutábilis, each of which was formerly considered as a distinct species. Further experience may, therefore, in due time, furnish the botanist also with some satisfactory test for the reduction of his species.

My own attention has been more particularly excited on this subject, by finding myself unexpectedly obliged to submit to the old opinion of Linnæus, in contradiction to that entertained by most modern botanists, that the primrose, oxlip, cowslip, and polyanthus are only varieties of one species. Upon what Linnæus founded his opinion, I know not; but, in vol. iv. p. 19. of the *Horticultural Transactions*, in a paper by the Hon. and Rev. W. Herbert, on the production of hybrids, there is recorded an experiment (which I see you have alluded to in your *Encyclopædia of Gardening*) so directly to the purpose, that no one who trusts to its accuracy can possibly resist its evidence. Mr. Herbert remarks : — " I raised, from the natural seed of one umbel of a highly manured red cowslip, a primrose, a cowslip, oxlips of the usual and other colours, a black polyanthus, a hose-in-hose cowslip, and a natural primrose bearing its flower on a polyanthus stalk. From the seed of that very hose-in-hose cowslip I have since raised a hose-in-hose primrose. I therefore consider all these to be only local varieties, depending upon soil and situation." I confess that I had myself given very little credit to this experiment of Mr. Herbert's, until it was recalled to my mind by a circumstance which I noticed in April, 1826, a few miles from Cambridge, at a place called Westhoe. I there found in great plenty a peculiar variety of Prímula, which I scarcely knew whether to call the oxlip or the cowslip. The leaves were larger and more downy than those usually found upon either of these plants; the flowers were in umbels, some drooping and others erect, and varying, in size and shape, from the ordinary character of the cowslip to that of the oxlip; the colour was as light a yellow as the usual tint of the primrose. Although this variety was every where abundant, both in the copses and open fields, neither myself nor a friend who was with me could find a single primrose in the neighbourhood, and comparatively few decided cowslips; which, however, were here and there scattered among this variety. At the very time that I first observed these plants, I was also much surprised at finding that a specimen of the cowslip, which had been transplanted into my garden when in flower the previous year, had completely changed its character this year; the limb of the corolla having become flatter and broader, the colour paler, and the whole appearance more like that of the oxlip. In the spring of the following year (1827), this plant threw up a few single-flowered scapes in addition to its umbels. The single flowers were somewhat smaller, and more deeply coloured, than those of the common primrose. After flowering the root decayed, and the plant separated into several smaller parts, which were transplanted, and have since flowered; each having both single and compound scapes. I may here observe, that plants with both single and compound scapes are very common in this neighbourhood, similar to that figured in the *Flòra Londinénsis*, which Dr. Hooker there styles the oxlip, and Sir Jas. Smith, in the *English Flòra*, considers to be a variety of the primrose.

I now collected the seed of some cowslips which were growing in a shady part of my garden, and sowed them in a similar situation. From this seed I have raised several plants, varying very considerably in their character. One is a perfect primrose; and all the rest approach, more or less, to the light-coloured variety of Westhoe. Not one has the decided character of the common cowslip.

I am not aware of any defect in the experiment from which this result has been obtained; but if the utmost accuracy be required by any one who may be still sceptical, I should recommend him to repeat the experiment, with the additional precaution of protecting the cowslips to be seeded from the approach of insects, that every possibility of the seedlings being hybrids may be avoided. The seed should be sown in a moist situation, and selected also from plants thus circumstanced; since it seems probable that such a combination of circumstances is better adapted to secure the developement of the characters of the primrose. The process which Mr. Herbert adopted, of highly manuring the cowslip from which his seed was selected, may possibly be still more likely to insure success.

If any of your correspondents can be prevailed upon to repeat these experiments in different parts of England, and favour you with the results, whether successful or not, they may do good service towards the final elucidation of this subject; and should Mr. Herbert ever present the public with a more detailed account of his own observations, he would confer an invaluable benefit upon those who wish to prosecute the enquiry. The best form for registering a series of such experiments, for the convenience of reference and comparison, would be, to class them numerically under different genera, and throw the remarks and occasional observations under separate heads, arranged as a table, being careful not to omit the mention of all the failures or unsatisfactory results.

I am, Sir, &c.

J. S. HENSLOW.

THE

MAGAZINE OF NATURAL HISTORY,

AND

JOURNAL

OF

ZOOLOGY, BOTANY, MINERALOGY, GEOLOGY,
AND METEOROLOGY.

VOL. III.
1830.

CONDUCTED

By J. C. LOUDON, F.L., G., & Z.S.

MEMBER OF VARIOUS NATURAL HISTORY SOCIETIES ON THE CONTINENT.

LONDON:

PRINTED FOR LONGMAN, REES, ORME, BROWN, AND GREEN,
PATERNOSTER-ROW.

1830.

ART. XI.   *On the Specific Identity of* Anagállis arvénsis *and* cærùlea. By the Rev. J. S. HENSLOW, Professor of Botany in the University of Cambridge.

Sir,

DR. HOOKER, in his most excellent *British Flora*, still keeps the *A*nagállis cærùlea distinct from the *A*. arvénsis; and perhaps the difference pointed out by him in the shape of the petals fully justifies him for so doing in the present uncertain state of our knowledge of the laws which regulate the limitation of species. I received last year, from the Reverend E. Wilson, some specimens and seeds of *A*nagállis cærùlea gathered in Yorkshire. From these seeds I have raised a dozen plants, nine of which have blue flowers, and three have red   Hence it should seem that in future *A*nagállis cærùlea must be considered as a variety of *A*. arvénsis. I conclude, from the above fact, that whatever may be the cause which disposes the petals of this species to assume a blue colour, this likewise disposes them to become notched or toothed at the margin, as Dr. Hooker has universally observed them to be. Though we cannot say the following law is certain in botany, yet it seems to me very likely to be true, viz. " That if a change takes place in one of the organs of a plant, a simultaneous change may be expected in some or all of the other organs considered to be modifications of the same organs." For instance, considering the calyx and corolla to be modifications of the leaf, when we see the leaf of the cowslip differing from that of the primrose, we need not be surprised to find that the calyx and corolla should differ also, though these plants be not distinct species, as I showed in my communication to the last Number of your Magazine.

I have met with the light pink variety of *A*. arvénsis at Higham, Kent; and the gentleman from whom I received the seeds of the blue variety mentioned in this communication, sent at the same time, and from the same place, some seeds of a white variety, similar to that received by Dr. Hooker from South Wales. From these seeds I have raised seven plants, one of which flowered red, and the other six white, tinged more or less with light pink, and having a bright pink eye. I wish I could persuade some of your correspondents to try similar experiments, in different parts of the kingdom, upon any of the various plants which approach each other so nearly as to leave it still a matter of doubt whether they ought to be considered distinct species or mere varieties of the same. I have some of them under trial, and propose sending you the results from time to time; but, as accidents are unavoidable, it would be more satisfactory to see them confirmed from different quarters.

I remain, Sir, &c.

Cambridge, September 17. 1830.   J. S. HENSLOW.

TRANSACTIONS

OF THE

CAMBRIDGE

PHILOSOPHICAL SOCIETY.

ESTABLISHED November 15, 1819.

VOLUME THE FOURTH.

CAMBRIDGE:

PRINTED BY J. SMITH, PRINTER TO THE UNIVERSITY:

AND SOLD BY J. & J. DEIGHTON, AND T. STEVENSON, CAMBRIDGE;
AND T. CADELL, STRAND, LONDON.

———

M.DCCC.XXXIII.

VIII. *On the Examination of a Hybrid Digitalis.*

By the Rev. J. S. HENSLOW, M. A.

PROFESSOR OF BOTANY, AND SECRETARY TO THE CAMBRIDGE
PHILOSOPHICAL SOCIETY.

[Read Nov. 14, 1831.]

ALTHOUGH the propagation of hybrid plants has been much attended to of late years by several Horticulturists in England, their experiments, for the most part, seem to have been undertaken for the sole object of encreasing the forms of beautiful flowers, or of modifying the flavour of delicious fruits. But the more curious and important physiological facts elicited by the phenomenon of hybrid productions do not appear to have received a proportionate degree of attention from those who have been engaged in these experiments. Chance having favoured me with a hybrid Digitalis during the past summer (1831), in my own garden, I employed myself, whilst it continued to flower, which was from June 19 to July 22, in daily examining its characters and anatomizing its parts of fructification. I was careful to compare my observations, with as much patience and accuracy as I can command, with the structure of its two parents. It seemed to me not unlikely that something interesting might result from a rigorous examination of this kind, or at least that its recorded details might serve as a point of departure for future observations.

The plant in question was undoubtedly a seedling from a specimen of D. lutea. I have this species and D. purpurea alone of the genus cultivated in my garden, where several plants of each had been allowed to scatter their seed, and the seedlings to grow wherever they chanced to come up. I had already remarked a singularity in the general appearance of one of these, and was watching the expansion of its flowers, when I was agreeably surprized to find it to be a decided hybrid, obviously having most of its characters exactly intermediate between those of purpurea and lutea. I had no doubt whatever of its being a seedling of lutea, from the position which it occupied in the garden in coming up amidst several plants of this species in a spot where an old plant had grown the year before; neither had any plant of purpurea grown in the same border Besides which, my plant exactly agrees in most particulars with a hybrid procured by Koelreuter in 1768 from seeds of lutea fertilized by the pollen of purpurea His account is accompanied by a rude and inaccurate figure which by no means tallies with his own description of the plant. In general habit, this hybrid approaches much nearer lutea than purpurea, Plate xv Fig 1. It is however decidedly taller and more robust than any specimens of the former species which my garden ever produced Koelreuter indeed asserts that the specimens raised by him were taller than either of their parents, but he assigns a lower limit to the height of purpurea than that to which many plants of this species have attained with me. Notwithstanding its more robust character and somewhat darker hue, the eye would scarcely have recognized, upon a mere casual observation and

---

* Acta Acad. Petropol. Anno 1777.

before its flowering, any peculiarity sufficiently striking to class it apart from some of the varieties of lutea, but a little closei inspection immediately detected certain decided points of difference. The whole plant is not so smooth as lutea, having a decided tendency to become downy, and being completely so on the under surface of the leaves, Plate xv. Fig. 2. The glabrous surface of lutea is one great characteristic of the species; though, if the D. rigida of Lindley is to be considered as a variety of it, which he seems to think probable, even this character fails. A few hairs are always indeed distributed here and there in the ordinary state of this plant, and seem to indicate the possibility of a transition from the one condition to the other, dependant probably on certain circumstances of soil or situation. From the ordinary condition of the leaves of lutea, however, those of the hybrid differ in a marked manner. They are even nearly as woolly on the under surface as the leaves of purpurea.

Examination of the external characters of the Hybrid.

I shall first describe the external characters of its several organs, comparing them with those of the parent plants. In Plate xvi, the corresponding parts in the fructification of the parents and of their hybrid are arranged in three columns, those of the latter occupying the middle column. A single glance of the eye will thus be sufficient to shew how exactly intermediate most of its organs are both in size and form, and in some cases also in color, to those of the two parents. There are however some remarkable deviations from this condition, which will be presently noticed.

Comparative view of the external characters of the three plants represented in PLATES XV. *and* XVI.

Purpurea.	Hybrida (purpureo-lutea)	Lutea.
	PLATE XV	
Biennial.	*Root.* Perennial, according to Koelreuter, and apparently so in the present instance, the plant having thrown out several offsets.	Bi-tri-ennial.
3—5 feet.	*Stem.* About 3½ feet.	2—3 feet.
1½—3 feet.	*Raceme.* About 1½ feet.	¾—1¼ feet.
less secund, and laxer.	secund, dense, nodding above.	denser.
woolly. very soft.	*Leaves.* Nearly smooth above, quite woolly below. Somewhat soft.	glabrous. firmer.
crenato-dentate.	Dentate.	dentate.
petiolate, oblong.	*radical,* sub-petiolate, broadly-lanceolate, Fig. 2.	somewhat narrower.
	caulinar, sessile, narrower.	
broader and shorter.	*Bracteas;* Lanceolate.	narrower & longer
longer than the Calyx and frequently than the bracteas.	*Pedicels.* About the length of the Calyx, and and somewhat shorter than the bracteas.	shorter than the Calyx and much shorter than the bracteas.
large, cernuous.	*Flowers,* medium size, nearly horizontal.	small, more drooping.
	PLATE XVI.	
I.	**II.**	**III.**
more spreading.	1. *Calyx,* moderately spreading in flower, afterwards connivent.	less spreading, at length more closed.
broader.	*a. sepals,* ovato-lanceolate, the odd one much narrower.	narrower.

If a general rule for naming Hybrids should be thought advisable, perhaps it will be found convenient always to prefix the name of the plant which supplies the pollen to that which furnishes the ovule.

Purpurea.	Hybrida (*purpureo-lutea.*)	Lutea.
more hairy.	*b.* hairy on the margins.	less hairy.
purple.	2. *Corolla,* Yellow ground tinted with red.	yellow.
spots more numerous, deep purple, and rings paler.	A few dark purplish-red spots surrounded by a paler ring in the throat and tube.	no spots.
less hairy.	Smooth, with hairs in the mouth.	more hairy.
obscurely 4 lobed, the upper emarginate.	Distinctly 4 lobed, the lobes blunt, the uppermost notched.	4 lobes deeper, acute, the upper deeply notched.
half the length, convergent.	3. *a. Stamens* length of the tube, nearly parallel.	somewhat more extended and divergent, according to Koelreuter; but I could see no very appreciable difference.
deeper orange-yellow, with numerous spots often confluent.	*b, c: Anthers* yellow inclining to orange, with a few small scattered purple spots.	lighter yellow, no spots.
much more oblique.	Oblique to the filament, converging above.	
	d, e: Pollen White, elliptic when dry, and sperical when moist. Some of the grains obscurely three-cornered, many are abortive, but those perfected are of exactly the same size and shape as in purpurea and lutea, being somewhat less than $\frac{1}{1150}$ of an inch in diameter.	
	4. *a. Pistil,* covered below with small glandular hairs.	
few hairs.	*style* cylindrical, with a few hairs on the lower part.	hair reaches higher up.
much more acute.	*b. stigma* cloven, very obtuse.	more acute.
more ovate and more pubescent.	*c. ovarium* oblong, pubescent.	more acute and less pubescent.
much more numerous.	*d. ovules* numerous, and exactly of the same shape and size as those of purpurea and lutea.	much less numerous.

Commentary on some parts of the preceding comparison.

Raceme. Although one of the characters of lutea lies in the very decidedly secund position of the flowers, some plants have them disposed in a squarrose manner round the axis.

1. *Calyx.* About one half the number of the flowers of the hybrid had five sepals and the other half six, (Plate xvi. II. 1. *c.*) and the sections given (from *d.* to *l.*) represent the different modes of their arrangement. Figs. *d.* and *h.* however appear to be their normal condition in æstivation, the other modifications having probably resulted from inequalities introduced during the expansion of the flower. The occasional development of a sixth sepal seems to be no uncommon occurrence in this genus, and I have met with it several times in specimens of lutea and ferruginea.

2. *Corolla.* In the colored copies of Professor Lindley's monograph, there are two varieties of lutea (see his Plates xxiv and xxv) in which the corolla is tinged with red. One of these (Plate xxiv) he considers to be a hybrid plant. In shape and size it approaches very nearly to the subject of the present paper, but the other (Plate xxv) more closely resembles lutea. In his figure of lutea also, (Plate xxiii) there is a little tinge of red in the mouth of the tube, on each side the base of the lip. I have never myself found the slightest tinge of red in any specimen of lutea, though the yellow is deeper and more inclining to orange in the parts above mentioned. If however it should be quite certain that genuine specimens of lutea do occur with a tinge of red in any part of their corolla, this circumstance must considerably modify our speculations as to how far the present hybrid may have derived this color from the male parent.

Flowers of lutea are not unfrequent with the lower lip notched (Fig. γ), which indicates the presence of a supernumerary petal blended into the tube of the corolla. In about half a dozen instances I even found this petal quite free, (Fig. β) and I believe occupying the same position as the sixth sepal in the anomalous cases just referred to In D. ferruginea, however, I have sometimes found a sixth sepal and a notched lip in the same flower. These anomalies may therefore be considered analogous phenomena among the supernumerary developments of the two organs.

3. *Pollen.* In comparing the action of the three pollens when immersed in water, I observed all the phenomena usually attendant on this experiment, to take place in those of purpurea and lutea: their grains quickly swelled and their granules were exploded in the form of a dense cloud (Fig. F and ζ). Two kinds of granules were also observed, the smallest and most numerous of which were too minute for me to be able to ascertain their precise shape and dimensions by the highest powers of my instruments; the others, much fewer in number, were considerably larger, and lay dispersed among the smaller like pellucid spots on a darker ground; and these might even be distinguished through the coats of the grains before their expulsion had taken place. Some pollen of purpurea taken from a withering stigma exhibited very distinctly the presence of the exserted membranous tubes (*boyaux*) described by A. Brogniart, Amici, and others, in the Ann. des Sciences, (Fig. G). Some of the granules also were marked on the surface by three blotches (Fig. H). Grains of pollen taken from the hybrid readily swelled upon immersion in water, though most of them appeared to be void of granules. Some few however certainly contained the larger

kind of granules, and I could see their explosion accompanied by successive and sudden contractions and dilatations of the grains themselves. But I could never detect any cloud of smaller granules similar to that which was exploded from the pollen of the parents, and which always proceeds from the grain by a continuous and slow emission, whereas the larger granules in the hybrid were discharged at intervals, and by separate efforts, and lay scattered at a distance from each other over the field of view (Fig. *f.*)

Koelreuter has given it as his decided opinion, derived from his numerous experiments, that true hybrids never reproduce their kind. Later experimenters have doubted this fact, and some seem to consider the question as quite settled to the contrary, at least with respect to the possibility of fertilizing a hybrid by the pollen of one or other of the parent species. But in prosecuting this enquiry we must be very cautious to keep in view the perfect distinctness of the two questions, whether it be *probable* and whether it be *possible* that hybrids should reproduce their kind. If it be *possible* that a true hybrid may do so, it may still be very *improbable*, from some deficiency in that connection of circumstances, of whatever description it be, which is essential to secure the fertilization of the ovule. We might imagine* for instance, so great a discrepancy to exist between the respective circumstances suited to the healthy action of its vegetative and reproductive functions, that although one climate may be adopted for securing the former, another might be required for obtaining

---

* This hypothesis is thrown out merely in the way of illustration, and not as likely to afford any solution of the cause of infertility observable in Hybrids, at least in most of them.

the latter, and thus the plant might continue to grow and flourish in one latitude, and yet be incapacitated for ripening its pollen or perfecting its ovules unless it could also thrive upon removal to another. There are certain plants, considered to be hybrids, which undoubtedly reproduce their kind freely enough; but some of these at least, if not all of them, are mere varieties of the same species. Thus Koelreuter ascertained that all the plants raised between D. purpurea and D. thapsi, by fertilizing the ovules of either by the pollen of the other, were constantly prolific, but then he also ascertained that D. thapsi itself when cultivated by him, after five generations assumed all the characters of purpurea. He consequently rightly inferred that D. thapsi was to be considered no otherwise than as a Spanish variety of the more common form of the species. If, again, it were possible for a true hybrid to be fertilized by the pollen of either of its parents, though it could produce no fertile pollen for itself, it would then evidently be in much the same condition as the female plant of any diœcious species, and its fertility might be secured by the instrumentality of insects, &c. In the present plant I repeatedly observed that the blossom always fell before the anthers on the shorter stamens had burst; and in order that this should not operate in diminishing the chance of impregnation, I touched some of the stigmas with the pollen extracted from these anthers, but without any success. Possibly however the pollen was not sufficiently ripened. I also touched other stigmas with the pollen of purpurea, and others again with that of lutea; but all these experiments failed in fertilizing any of the ovules. Koelreuter was equally unsuccessful in his attempts to fertilize this hybrid. I must here record what has appeared to me a remarkable circumstance, brought before my notice during the prosecution of these enquiries. There were three or four plants of *lutea* in my garden which were quite deficient in pollen, and which nevertheless produced perfect seeds. I was unable to detect even a single grain of pollen either healthy or abortive in their anthers, though these latter organs appeared to be well formed and perfected. The ovaria of these plants indeed contained plenty of ovules, most of which I afterwards observed had been fertilized, since their seeds ripened. These plants must therefore have been fertilized by the pollen of other specimens in their neighbourhood; at least according to all our present notions on this subject. But then the ovules of the hybrid were also similarly circumstanced, and if they had been capable of receiving the same influence from other plants, there is no apparent reason why they should not have proved fertile also.

4. *Ovules.* In the parent plants, the ovules begin to grow and develop themselves immediately after the fall of the corolla, whilst in the hybrid they soon wither away. It is remarkable however, that all symptoms of decay in the ovarium are strictly limited to the ovules themselves, for even the little protuberances upon which they are seated on the placenta remain succulent, as do the various parts of the pericarp, including also the base of the style: all which continue healthy and attain their perfect dimensions, the valves alone slightly collapsing from the deficiency of the ovules in the enlarged cells. Plate XVII. Fig. 4. But the stigmatic tissue dries up, and a cavity is thus left through the upper part of the dissepiment, forming an opening between the two cells, Fig. 5. *e.* The same effect sooner or later takes place also in the seed vessels of the parents.

Recapitulation In reflecting upon the points of resemblance and of disagreement in the organs of fructification of these three plants, the most striking circumstance which we have hitherto noticed in their external characters, is the perfect identity in size and shape both of their pollen and of their ovules. As the respective organs which contain these bodies, viz. the anthers and the ovaria, are each proportionate to the different sizes of the three flowers themselves, it is evident that a flower of lutea must have much less pollen and many fewer ovules than one of purpurea, which in fact the most casual observation is sufficient to shew. The ovules of the hybrid also are about intermediate in number to those produced by the parents. It will be a subject worthy of future investigation, to determine whether one condition necessary for securing the hybridity of two species, require their pollen and ovules to be of the same, or of nearly the same dimensions. Except in the above instances, and in the very peculiar shape of the stigma, all the other external characters of the hybrid appear to be precisely intermediate between those of its parents. The chief physiological difference observable in the external economy of the organs of fructification seems to reside in the fall of the corolla, which in the parents does not take place till after the anthers have discharged their pollen and become perfectly withered, whereas in the hybrid the corolla falls before the anthers on the shorter stamens have burst, and when even those on the longer pair, although opened, have hardly parted with their pollen, and have not as yet become in the least withered. The style and stigma of all three appeared to comport themselves alike, that is to say, they all began to wither soon after the fall of the corolla.

Examination of the internal structure of the Organs of Fructification.

Before I begin the detail of this examination, I may at once state, that so far as I have hitherto been enabled to pursue it, I have not perceived the slightest difference between the internal structures of the three plants; and as their organization is somewhat different from any of the cases selected by Mons. A. Brogniart to illustrate his paper on the formation and developement of the embryo, the present attempt may not be without some general interest to the physiologist, independent of the objects connected with the particular enquiry for which it has been undertaken. The method which I pursued was always to examine the various parts dissected, first, in specimens of purpurea, and then to comcompare them with the like parts in hybrida, and lutea. Though it is possible therefore that I may accidentally have overlooked some defect and dissimilarity in the internal structure of the hybrid during this common and simultaneous examination of all the three, and may have represented in the drawings some appearance or other strictly belonging only to the anatomy of purpurea, yet I do not think such an error could very probably have occurred. As the main object in view was the direct comparison of the three plants, any striking difference at least would have been noticed, and the subject have been submitted to a rigorous re-examination.

Vessels of the Pistil. Plate XVII. Fig. 1. represents a longitudinal section of the ovarium perpendicular to the dissepiment, and consequently passing through both the cells; and Fig. 2. is another longitudinal section, at right angles to the last, and through the plane of the dissepiment, or rather, it represents the surface obtained by tearing the ovarium asunder down the thickness of the dissepiment, which is composed of two skins with parenchymatous matter between them. The threads of vascular tissue arranged in a circle round the axis of the pedicel (*a*), after giving off veins to the calyx and corolla (*b*), and again to the pericarp (*c*), diverge on either side into the placenta (*d*), a little above its lowest point, and then ramify or subdivide through its substance into separate fibres (*d'*) which proceed directly to the bases of the ovules. Fig. 3. represents a transverse section of the upper part of the ovarium with the lower part of the style; the valve which is nearest the spectator being removed, as also are the ovules in this cell. The smaller veins (*c'*), of which more than twenty are seen rising through the pericarp, all terminate in the base of the style; but the two larger ones (*c*), which run along the loculicidal edge of the pericarp, rise through the whole length of the style. The stigmatic tissue (*e*), (Fig. 1. 2. 3.) descends down the middle of the style till it comes into contact with the summit of the placenta. When the appearances here represented are examined with the highest magnifiers, their more intimate structure is exposed, as in Plate XVIII. where Fig. 1. and 2. are two transverse sections of the pistil, of which the former corresponds to one quarter of the circumference of the ovarium represented in the lower part of Fig. 3. Plate XVII., and the latter agrees with the section through the style in the upper part of the same figure. Plate XVIII. Fig. 3. and 4. are longitudinal sections of the same organ, the former through the stigma, the latter through the summit of the ovarium where the stigmatic tissue (*e*) descends to the placenta, as in Fig. 1. Plate XVII. In these highly magnified sections all the corresponding parts are designated by the same letters as in the former figures.

The veins (c), (d), &c. are in all cases composed of bundles of tracheæ, which in the larger veins (c) are very numerous. I have counted sometimes between thirty and fifty combined in the construction of a single vein (c), a fact which would not be suspected upon a casual observation, but which becomes evident by digesting the style in nitric acid, when these elementary parts are easily separated. Their terminations are in the form of elongated cones, and they all end together, a short distance below the stigma. (See Plate xviii. Fig. 3.) The other elementary parts of all these veins are certain extremely delicate tubes which invest the central bundle of tracheæ, and give it the appearance of being surrounded by a mucous or glutinous substance, but which under the highest powers of the microscope may be separated into these tubular vessels, whether subdivided or not by transverse diaphragms, I was unable to satisfy myself. This very delicate tissue has the same general appearance as the stigmatic tissue, which in these plants descends down the centre of the style, to the summit of the placenta. Where this latter tissue terminates in the stigma, it is indeed evidently composed of distinct cells, easily separable from each other by nitric acid, Plate xviii. Fig. 3. (q). Lower down however the cells are more elongated (r), and lower still, where this tissue meets the placenta, I could neither detect any transverse diaphragms in it, nor even detach its cells (if they were such) from each other at their extremities by the action of nitric acid, though they were easily separated longitudinally into long filamentous strings. In this part of its course therefore the stigmatic tissue appears rather to be tubular than cellular in its structure. After this tissue has become divided into two bands, penetrating on either side through the dissepiment into the two cells, it seemed to me, upon a most careful examination, to coat over the whole *surface* of the placenta. It is very difficult however to be quite certain of this fact, and I may be wrong; but after numerous dissections made upon the three plants, I found I could generally raise, with the point of a very fine needle, a thin gelatinous film of a delicate fibrous structure from between the ovules Fig. 4. (*e'*), which film seemed to be similarly constituted, and also continuous with the stigmatic tissue (*e*).

Cellular tissue of the Pistil. These cells are for the most part compressed into tolerably regular rhomboidal dodecahedrons, excepting in the placenta, where, as the ovarium increases, the vesicles assume that irregular character so well described and represented by Mons. A. Brogniart in the parenchyma of the leaf, (Ann. des Sc. Vol. xxi.) and they have the same sort of interstices filled with air between them as those which occur in that organ. When the style is digested in nitric acid, the separate vesicles of its cellular tissue become cylindric-oval, Fig. 5. (*o*): and I have represented an appearance (*p*) which was noticed several times upon some of these vesicles, of a faintly marked band running down one side. — Further examination may perhaps throw some additional light upon this circumstance, but at present I know not to what cause it may be ascribed.

Epidermis of the Floral Organs. Plate xvii. Fig. 6, 7. The flattened cells are of the same size in the three plants, their diameter being somewhat more than the thousandth of an inch. They vary in shape from hexagonal to quadrangular prisms bordered by straight, or waved sides. This membrane is irregularly supplied with stomata (*f*). When digested in nitric acid, the cells assume an appearance represented in Fig. 7., as though the granular matter they contain were coagulated into a nucleus, or else were enclosed in a separate internal vesicle. Whether this appearance originate in any optical deception, I could not sufficiently satisfy myself; but if, as I am inclined to think, it does not, the fact must have been hitherto overlooked from the difficulty of detecting the true plane of junction between the contiguous cells, owing to the very great transparency of their membrane. Thus, in Fig. 6, where this epidermis is less magnified, the cells appear to be separated from each other by anastomosing veins or canals, whilst in Fig. 7. it is shewn that their true planes of junction run directly along the middle of these canals. I am however quite positive upon another point which has been a subject of dispute among physiologists; I mean the existence of a delicate homogeneous membrane investing this epidermis. Such a membrane may be distinctly separated by the action of nitric acid, from the epidermis of the corolla, filament, and style. It is faintly marked by parallel longitudinal striæ Fig. 7, (*g*), and appears to coat over the whole surface of these organs, but whether it is perforated by a fissure opposite each stoma I did ot ascertain.

Structure of the Filament. Plate XVII. Fig. 8, 9. The cellular tissue of this organ consists of elongated rhomboidal dodecahedrons, as the elongated hexagons seen in its longitudinal section sufficiently explain (Fig. 9.). A single bundle of tracheæ runs up the middle of it, invested by the peculiarly delicate fibrous tissue already noticed.

Structure of the Anthers. Plate XVII. Fig. 10—12. The fibrous cells* composing the inner coat of the anther, appeared to me quite as distinct and perfect in the hybrid as in the parents. Nor did I observe the slightest difference in the formation and

---

* See Purkinje " De cellulis antherarum fibrosis, &c. 4to. Vratislaviæ 1830."

condition of any part of this organ in either of the three plants. In general, a transverse section shewed the fibrous-cells to be arranged in a triple tier (Fig. 10.). These curious vessels seemed to be set, as it were, upon the sides and edges of void dodecahedral and other polyhedral spaces, as though certain original cells of these shapes had disappeared and left this frame-work of their structure alone standing. The triple tier is not distinguishable upon looking directly down upon the inner surface of the anther (Fig. 11.), but some of the fibrous-cells may be seen standing upon the junction-edges of the cells of the epidermis, where this membrane has been partially cleaned of the inner coating composed of them. Fig. 12. (*h*) is the appearance which they assume when detached by digestion in nitric acid: (*k*) being the cells of the epidermis, (*l*) an accidental appearance in a grain of pollen recalling somewhat of the character of the grain figured at Plate xvi. Fig. 3. H.

Structure of the Ovules. Plate xvii. Fig. 13. When the corolla is expanded, the ovules are entirely composed of a con-geries of large vesicles, and their surface has a very remarkable and granulated appearance. At this period of their existence I was unable to detect any thing very precise respecting the distinction and distribution of their several parts. The fora-men (*m*) however was evidently seated near the hilum, and a darker spot indicated the chalaze (*n*) to be at the opposite ex-tremity (see also Plate xviii. Figs. 1. and 4.) In the ovules of purpurea and lutea, there is no difficulty in tracing the separate parts of the ordinary structure, if they be examined shortly after their impregnation; but before their fertility is secured I have not hitherto been able to detect in these plants, more than in the hybrid, any thing but a homogeneous mass of cellular tissue.

Possibly I have not given this part of .the investigation sufficient attention. When the ovules are digested in nitric acid, the detached cells assume an oval shape, Fig. 15. (*o*), and are yellowish. But among them I several times observed a larger cell (*p*) which was more transparent and whiter, and which I fancied might be the origin of the embryonic sack. These component parts are best exhibited by crushing the ovule between two flat pieces of glass. Fig. 14. represents a monstrosity in which an ovule was observed to stand upon a sort of pedicel.

Recapitulation. So far then as these researches have hitherto proceeded in comparing the internal structure of the floral organs of the hybrid with those of its parents, no appreciable difference has been detected. The elementary vesicles of which their cellular tissue is constructed seem to be all of the same size, and consequently it is evident that fewer of these vesicles must be employed in the conformation of any of the parts of hybrida, and still fewer in those of lutea, than in completing the corresponding parts of purpurea. But there appears to be nothing actually defective in any part of these organs in the hybrid, nothing wanting of whatever is to be found in those of the two parents. The nutritive apparatus more especially, so far as we have examined it, seems to be quite perfect, and as the functions performed by it in all three plants are precisely the same up to the period when the flower falls, there seems to be no reason for suspecting the hybrid to differ in any particular from its parents in the perfection of its conservative organs. Since however the functions of the reproductive apparatus appear to cease in the hybrid before they do in the parents, it should seem that there must be some deficiency in this part of its organization, though it has not yet been noticed. Should the Society consider the details of this examination worthy their attention, I propose to myself the further satisfaction of prosecuting it afresh next summer, if another opportunity should be permitted me. Indeed I ought to add, that in the present state of this enquiry, so little additional light has been thrown upon the great questions connected with the phenomenon of hybridity, that I should hardly have felt myself justified in presenting these remarks to their notice, were it not in the hope that they might save some time and trouble to whomsoever may be inclined to take up the subject, and possess the means of carrying on the investigation of it still further.

DESCRIPTION OF THE PLATES.

PLATE XV.

THE raceme (Fig. 1.) and radical leaf (Fig. 2.) of the Hybrid.

PLATE XVI.

The various parts of the floral organs in the three plants contrasted together.   The details are at pages 4 and 5.

> As the same parts in the three columns are marked by corresponding letters in three alphabets, viz. in Roman capital, small Italic, and Greek characters, it will be unnecessary to refer to more than the figures in one compartment for the purpose of explaining those in the others.

1. *Calyx.*   A. sepals separated and spread open: B. their marginal hairs magnified: *c.* with supernumerary sepal: *d.* to *l.*, arrangement of the sepals during inflorescence. N. B. These sections do not refer to the arrangement of the sepals in *æstivation,* which by some neglect I omitted to notice.

2. *Corolla.*

   β. with supernumerary petal: γ. ditto blended with the tube and forming a notched lower lip.

3. *Male Organs.*   A. is of the natural size; the rest are more or less magnified.

   A. Position of the stamens in the tube of the corolla: B. a front, and C. a back view of the anthers: D. dry, and E. moistened grains of pollen, lying on squares representing the $\frac{1}{1150}$ of an inch: F. a grain exploding upon the application of moisture: G. three grains taken from off the surface of a withering stigma, with their tubes (*boyaux*) exserted: H. a grain with three lighter blotches on the surface.

4. *Female Organs.*   A. is of the natural size; the rest are more or less magnified.

   A. pistil: B. stigma: C. transverse section of the ovarium: D. an ovule at the period of the flowers expansion, placed on a micrometer divided to the $\frac{1}{500}$ of an inch.

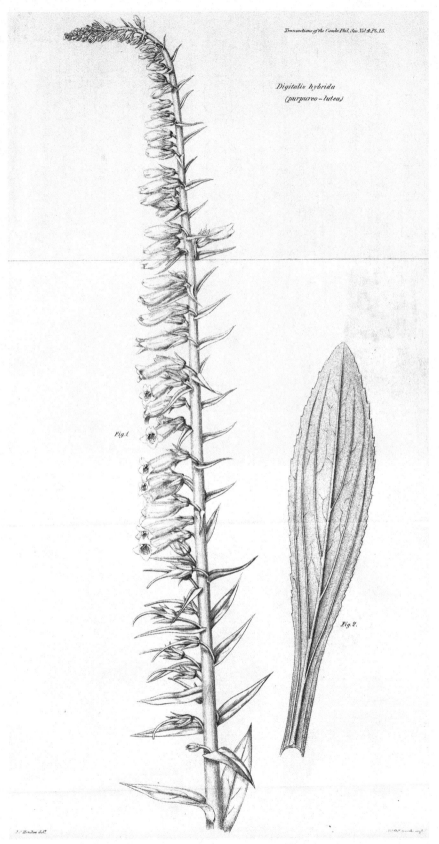

Digitalis hybrida
(purpureo – lutea)

Fig.1.

Fig.2.

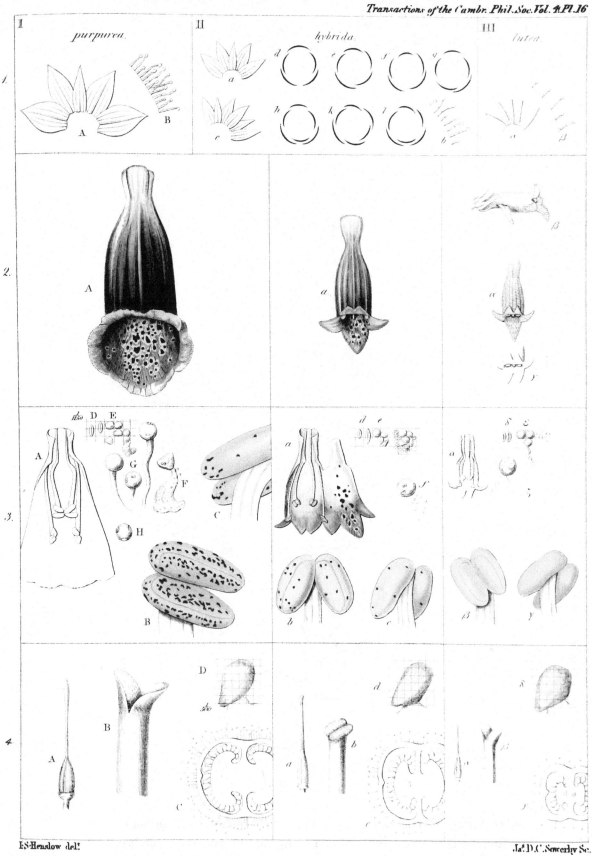

I.S. Henslow del.

Ja.^s D.C. Sowerby Sc.

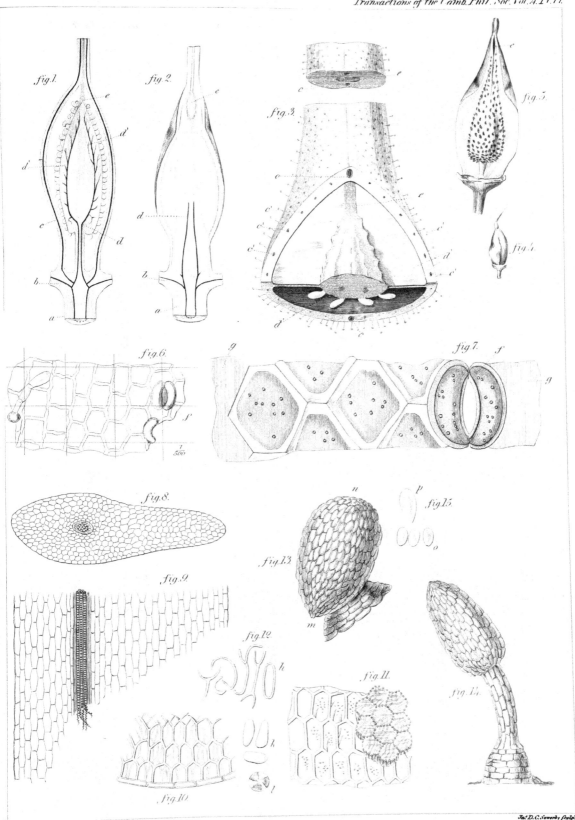

L.S. Henslow delt.

Jas. D.C. Sowerby sculp.

Plate XVII.

Anatomy of the parts of fructification. All the figures excepting Fig. 4, are more or less magnified. The same letter is always employed to designate the same parts in the different figures.

Fig. 1, 2. *Ovarium*, longitudinally divided; in the first case perpendicular to, and in the second down the plane of the dissepiment.

a. The pedicel with its circle of vascular bundles surrounding the axis: *b,* branches of this circle given off to the calyx and corolla: *c,* two larger bundles which run up the pericarp, along the future line of its dehiscence, and rise through the whole length of the style: *d,* separation of the vascular bundles into two bands which enter the two lobes of the placenta near their base, and rising through their substance *d`,* again separate and subdivide, giving off single vessels to the bases of the ovules: *e,* the stigmatic tissue descending through the style to the summit of the placenta.

Fig. 3. A transverse section through the summit of the ovarium, and again through the base of the style. The valve and ovules of one cell are removed. The letters designate the same parts as in the last Figure, with the additions of *c`,* small vascular bundles rising through the pericarp, all of them terminating in the base of the style.

Fig. 4. Ripened pericarp of the Hybrid, of the natural size.

Fig. 5. The same magnified, with one valve removed—exhibiting the dissepiment, and one lobe of the placenta, which is still fleshy, and covered by abortive ovules: *e* a cavity left by the drying up of the stigmatic tissue.

Fig. 6. *Epidermis* of the corolla, with a glandular hair and two stomata (*f*).

Fig. 7. The same digested in nitric acid and more highly magnified; *g,* being the investing pellicle faintly but very regularly striated.

Fig. 8, 9. *Filament;* transverse and longitudinal sections.

Fig. 10. *Anther;* a section perpendicular to its coats, exhibiting the triple tier of its fibrous cells.

Fig. 11. A fragment of the coats of the anther viewed on the inside perpendicularly to its surface, which is partly divested of the fibrous-cells.

Fig. 12. Details of the anther after it has been digested in nitric acid; *h,* fibrous-cells *k,* vesicles of the epidermis; *l,* a grain of pollen peculiarly marked.

Fig. 13.   *Ovule ; m,* foramen ; *n,* chalaze.

Fig. 14.   Monstrosity of ditto.

Fig. 15.   Details of the ovule after digestion in nitric acid ; *o,* the smaller vesicles composing the bulk of the ovule ; *p,* a paler colored vesicle occasionally found among the former.

PLATE XVIII.

Highly magnified sections of the style and ovarium.   Wherever the same letters are used in this plate as in the last they designate the same parts.

Fig. 1.   Transverse section of one quarter of the upper part of the ovarium.   Fig. 2. Transverse section of the style.   Fig. 3. Longitudinal section of the stigma, and part of the style.   Fig. 4. Longitudinal section of the base of the style and apex of the ovarium, perpendicular to the plane of the dissepiment.

c, the two large veins, or bundles of tracheæ, which rise through the whole length of the style: *c',* the numerous smaller veins which terminate in its base: *d',* fragments of the vascular bundles which rise into the placenta and branch off to the ovules : *e,* stigmatic tissue descending down the centre of the style to the summit of the placenta ; *e',* the same tissue coating over the *surface* of the placenta, and passing round the bases of the ovules : *m,* foramen, and *n,* chalaze, indicated by darker spots ; and in Fig. 1. the position of a raphe is apparent through the ovule, by a darker band extending from the hilum to the chalaze : *q,* vesicle of the stigma : *r,* tubular vesicles of the stigmatic tissue.

Fig. 5. *o,* Vesicles of the cellular tissue of the style detached by digestion in nitric acid : *p,* one of them marked by a transverse band, when seen more highly magnified.

THE

MAGAZINE OF NATURAL HISTORY

AND

JOURNAL

OF

ZOOLOGY, BOTANY, MINERALOGY, GEOLOGY, AND METEOROLOGY.

CONDUCTED

By J. C. LOUDON, F.L., G., & Z.S.

MEMBER OF VARIOUS NATURAL HISTORY SOCIETIES ON THE CONTINENT.

LONDON:

PRINTED FOR

LONGMAN, REES, ORME, BROWN, GREEN, AND LONGMAN,
PATERNOSTER-ROW.

1832.

The flowering stems of Pàris quadrifòlia (*fig.* 86. *a*) bear one whorl of leaves, and four whorls in the floral organs; and in the most common state of the plant these whorls are respectively composed of four leaves, four sepals, four petals,

86

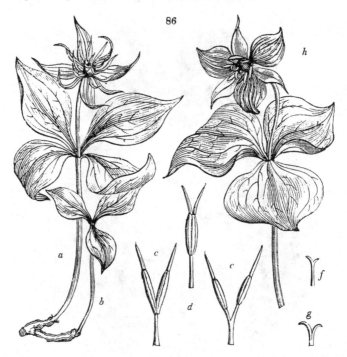

eight stamens, and a four-celled pistil, crowned by four stigmas. In this state, therefore, it offers a marked exception to the law which is so prevalent among monocotyledons, " that the number 3, or a multiple of it, should prevail in the developement of some part or other of their structure."

The frequency, however, with which this plant deviates from its more common condition, seems to indicate a great degree of instability in the operation of whatever be the law which regulates the developement of its subordinate parts; which should make us cautious in pronouncing upon the normal condition of its several foliaceous whorls. It seems to me that some light may be thrown upon this question by examining a great number of specimens from different localities, and recording the *limits* within which the number of parts in each organ is found to vary. With this view, I have, for the last three or four years, noted the different varieties gathered by myself and two friends, Messrs. Babington and Downes, in a habitat near Cambridge. The result of our examinations, made upon 1500 specimens, I have arranged in the following tables, upon which I shall offer a few remarks.

TABLE I. — The condition and number of each, of 38 distinct varieties observed among 1500 specimens.

Variety.	Leaves.	Sepals.	Petals.	Stamens.	Stigmas.	Number of instances of each variety.	Variety.	Leaves.	Sepals.	Petals.	Stamens.	Stigmas.	Number of instances of each variety.
1	3	5	3	8*	4	1	21	5	5	4	8	4	1
2	4	3	3	6	3	1	22 {	5	5	4	9	4	10
3	4	3	3	7	3	2		5	5	4	9	4*	1
4 {	4	4	3	7	4	3	23	5	5	4	9	5	1
	4	4	3	7*	4	1	24	5	5	4	9	6	1
5	4	4	3	8	3	2	25	5	5	4	10	4	1
6	4	4	3	8	4	2	26	5	5	4	10	5	1
7	4	4	3	9	4	1	27	5	5	4	11	5	1
8	4	4	4	8	3	4	28	5	5	5	10	5	6
9 {	4	4	4	8	4	1160	29 {	6	3	3	7	4	1
	4	4	4	8*	4	3		6	3	3	7*	4	1
	4	4	4	8	4*	1	30	6	3	3	8*	4	1
10	4	4	4	8	5	12	31	6	4	3	8	4	1
11	4	4	4	9	4	19	32	6	4	4	8	3	1
12	5	3	3	7	3	1	33	6	4	4	8	4	1
13	5	3	3	7	4	1	34 {	6	4	4	8	4	12
14	5	4	3	7	4	2		6	4	4	8*	4	2
15	5	4	3	8	4	2		6	4	4	8	4*	1
16 {	5	4	4	8	4	192	35	6	4	4	9	4	4
	5	4	4	8*	4	2	36	6	4	4	10	4	1
17	5	4	4	9	4	31	37 {	6	4	4	9	4	1
18	5	4	4	9	5	1		6	5	4	9	4*	1
19	5	4	4	10	4	3	38	6	5	4	10	4	1
20	5	5	3	8	5	1							

TABLE II. — Exhibiting the number of times that the several variations in each whorl occur throughout the whole number of specimens.

Num. of parts.	3	4	5	6	7	8	9	10	11	12
Leaves -	1	1211	259	29						
Sepals - -	8	1464	28	0						
Petals - -	24	1470	6	0						
Stamens -	—	—	—	1	12	1402	71	13	1	0
Stigmas -	11	1465	23	1						

TABLE III. — The numerical proportion between the most common condition of the several whorls (as in No. 9.) and the other cases, in which their several parts are either increased or diminished.

	Leaves.	Sepals.	Petals.	Stamens.	Stigmas.
Parts diminished	1	8	24	13	11
Ordinary state -	1211	1465	1470	1402	1465
Parts increased	288	28	6	85	24

In Table I. an asterisk (*) is placed against the number of

the stamens and stigmas in some subvarieties, in which one of these organs exhibited a tendency to subdivide, or become double : by a filament bearing two anthers (*fig.* 86. *c, d, e*) or by a stigma becoming branched (*f* and *g*). By this table it appears that the most common variety (No. 9) exceeds three fourths of the whole number of the specimens examined ; and that No. 16, next in abundance, which differs from it only in having one leaf more, comprises more than half of the remainder. Together, these two varieties exceed nine tenths of the whole.

By Table II. we are shown the *limits*, within which the number of parts developed in the separate whorls may lie : and it is very remarkable that these limits are so nearly 3 and 6 for the four whorls of leaves, sepals, petals, and stigmas ; and that they are nearly 6 and 12 for the whorl of stamens. There are, in fact, only three varieties wanting out of the twenty-three which such conditions would render possible, and these deficiencies are confined to the non-occurrence of the highest limits assumed for the sepals, petals, and stamens. It should be remarked, also, that although a single instance only has occurred, in any of the flowering stems, in which the first whorl was composed of three leaves, this number, however, is very general in the foliaceous whorl that crowns the barren stalks (*b*).

Table III. is only a slight modification of Table II., but shows us a little more clearly what is the tendency of each whorl to deviate, by excess or default, from its most common condition.

Supposing, now, by way of hypothesis, we assume the normal character of the plant to be, that it have its several whorls composed of the minimum values which have been observed in the numbering of their separate parts, we should reduce it to the state of a Tríllium (*h*), an allied genus, which affords as excellent an example of the law of subdivision prevalent among monocotyledons*, as the usual state of Pàris quadrifòlia is a remarkable exception to it. If we would now attempt to account for the reason of such an ano-

* " That the number 3, or a multiple of it, should prevail in the developement of some part or other of their structure." In the genus Tríllium, as the term trillium implies, the parts are all triple ; the leaves are 3 ; the sepals (leaflets of the calyx) 3 ; the petals 3 ; the stamens 6, that is, twice 3 ; the stigmas 3 ; and the cells of the berry 3. (See *fig.* 86. *h* above.) The primordial veins, too, of the leaves, sepals, and petals, both of the species of Tríllium and of Pàris quadrifòlia, are usually three. The species of Tríllium are all natives of America, where more than thirty species, it is said, have been discovered; out of this number thirteen species have been introduced into British gardens. The figure above (86. *h*) represents Tríllium eréctum *L.*, and is copied from t. 470. of Curtis's *Botanical Magazine.* — *J. D.*

maly, we may suppose that this plant is ever struggling, as it were, to become double in all its parts; but that it seldom succeeds, except in the case of the leaves, in subdividing and developing any more than *one* of the subordinate parts of each separate whorl. If each part were split up into two, by the sort of process exhibited in *fig.* 86. *c* to *g*, the plant would then acquire the maximum of developement indicated by the law suggested from Table II. But if, on the other hand, we assume that these maximum values belong to the normal condition of this Pàris, we have still a monocotyledon regularly subdivided into multiples of 3 ; only now we must ascribe its ordinary character to a constant tendency to abortion in the separate parts of each whorl. Which of these two hypotheses, or whether either of them, may be correct, it would be premature to decide. I shall, however, be very happy in finding any of your correspondents inclined to assist me in the investigation, by constructing similar tables from specimens procured in different habitats. I would, however, suggest an improvement in the mode of making these observations, which did not occur to me before, viz. to estimate the number of parts in the innermost whorl (or pistil of the flower), from the number of *cells* in the ovarium, and not from the number of the stigmas. Whenever there are more stigmas than cells, an asterisk may then be placed against the subvariety, as in Table I., indicating a tendency in some part of this whorl to become double.　　　　　I am, Sir, yours, &c.

Cambridge, Feb. 4. 1832.　　　　　　　　J. S. Henslow.

In Vol. IV. p. 446-7. is a list of the rarer plants of Essex; and J. G., its author, remarks : — " In the Thrift Wood, near Chelmsford, Pàris quadrifòlia thickly covers the sloping sides of a pond (which is filled with Hottònia palústris), and grows to an unusual size : many of the specimens have five leaves." This remark suggests some connection between a vigorous condition of the plant, and the production of a fifth leaf.

Gerard Edwards Smith (of St. John's College, Oxford), in his *Catalogue of the Phænogamous Plants of South Kent*, states that he met with several specimens of Pàris quadrifòlia, precisely in the condition of Professor Henslow's twenty-eighth variety above, in a wood at Stowting; and adds that he found such specimens to be severally furnished with a 5-celled seed-vessel. Of one of these specimens he figures, in plate i of his *Catalogue*, a flower, to exhibit the quinary division of its parts, and gives beside it a detached figure of the 5-angled 5-stigmaed germen, and another of a transverse section of it, for the sake of displaying the 5 cells of which it consists, and that one of these cells is larger than the remaining 4, as it is remarked to have been in the explanatory description.

Sir J. E. Smith, in Rees's *Cyclopædia*, describes two species of Pàris; one, our English P. quadrifòlia; the other, a species from Nepal, which he denominates P. polyphýlla. His entire description of it is here presented : — " P. polyphýlla, Many-leaved herb Paris. Leaves lanceolate, 8 or 10,

Discovered by Dr. F. Buchanan, growing in woods near rivulets, in Upper Nepal, where it is known by the name of Dai Swa. The root is creeping, but much thicker and more tuberous than in P. quadrifòlia. Stem a foot or more in height, brownish, thrice the thickness of that of P quadrifòlia, crowned like that with a whorl of leaves, in a similar spreading position, but about twice as numerous, and much narrower, being elliptic-lanceolate, taper-pointed, triple-ribbed; each supported on a short purplish stalk. Flower-stalk much shorter than that of P. quadrifòlia, but the flower is larger, of the same green colour. Its calyx leaves and petals are usually 5, with ten stamens; but sometimes only 4 with 8 stamens; or even 3 with 6. The styles, as well as the cells of the germen, always agree in number with the petals; but the former are combined by a thick columnar base, which character, added to his not having seen the ripe fruit, caused Dr. Buchanan to doubt of the genus. The habit and rest of the characters, however, leave no scruple in our mind; but it is very possible that what Linnæus, in the European species, calls styles, may be almost wholly stigmas, for they are downy all along their upper side, and the Nepal plant will then be found to differ merely in having the style more elongated."

Sprengel, in his *Systema Vegetabilium* (vol. ii. p. 261.), registers the following species of Pàris : —

" 1. P. quadrifòlia *Linn.* Leaves 4, in a whorl, sepals linear, exceeding the petals in length. Inhabits the shady woods of Europe.

" 2. P. verticillàta *Bieberstein.* Leaves 8, in a whorl, sepals lanceolate, thrice as long as the petals. Inhabits the east of Siberia and Nepal. [Synonyme:] P. polyphýlla of *Smith.*

" 3. P. incomplèta *Bieberstein.* Leaves about 10, in a whorl, sepals lanceolate, nerved; petals none; anthers not lengthened out at their tips. Inhabits Armenia and Iberia. [Synonymes:] P. apétala and Demidòvia polyphýlla [both of] *Hoffmann.*"

This quotation from Sprengel informs us of the existence of at least two species of Pàris, which individually produce a whorl of about 9 leaves; and should Sprengel, in his haste, have wrongly identified Bieberstein's P. verticillàta from the east of Siberia, with Smith's P. polyphýlla from Nepal, then three species will be known which severally produce a whorl of about nine leaves.

Has any one, by night or by day, observed the flower of Pàris quadrifòlia to be fragrant? Its green colour suggests the likelihood of its being fragrant, as does the fact that the flower of Tríllium díscolor *Wray*, a species native to Georgia in America, and figured in Curtis's *Botanical*

Magazine, t. 3097., is described to exhale an odour resembling that of the blossoms of the American allspice (Calycánthus flóridus). These are fragrant, indeed; for when numerously expanded, in the sunny days of July, their usual time of flowering, they diffuse a copious volume of delicious and aromatic odour, capable of perfuming the surrounding air for many yards. — *J. D.*

THE

FOREIGN

QUARTERLY REVIEW.

VOL. XI.

PUBLISHED IN

JANUARY AND APRIL,

M. DCCC. XXXIII.

LONDON:

TREUTTEL and WÜRTZ, and RICHTER,

30, SOHO SQUARE:

BLACK, YOUNG, AND YOUNG,

TAVISTOCK STREET.

1833.

ART. III.—*Physiologie Végétale, ou Exposition des Forces et des Fonctions vitales des Végétaux, pour servir de suite à l'Organographie Végétale, et d'Introduction à la Botanique Géographique et Agricole.* Par M. Aug. Pyr. De Candolle. 3 tom. 8vo. Paris. 1832.

THE great importance of vegetable physiology is sufficiently evident. The agriculturist and the horticulturist can expect increased success in their several departments, in proportion only as their practice reposes on an improving knowledge of the laws which regulate the phenomena of vegetable life. We have long wanted a work in which the more recent discoveries of modern observers should be collected, and their facts generalized; and the present volumes will be found to supply this want in a very efficient manner. The great progress which has been made of late years in this subject, could be known only from consulting the papers of various contributors, scattered through the pages of different scientific journals; and there existed no complete treatise to which the botanist might refer for an extensive and combined view of the several laws and principles which had been either clearly established or strongly suggested by a closer examination of the complicated phenomena of vegetation than had previously been attempted. On the continent, a long list of names might be enumerated of those who have devoted their attention to vegetable physiology; but in England, with a few rare exceptions, our best botanists have suffered their continental brethren to outstrip them in this superior department of the science. Whilst we possess at least a sufficient number of works exclusively devoted to " descriptive botany," we can scarcely name one that makes any pretension to a close acquaintance with the more recent discoveries in " vegetable physiology."* Mrs. Marcet's little work, entitled, " Conversations on Vegetable Physiology," is, indeed, excellent of its kind, and may be read with advantage and pleasure by every one who wishes to obtain a superficial knowledge of the subject. It professes merely to give an exposition of some of the leading topics of M. De Candolle's lectures, in his annual course at Geneva. We have now, however, the views of De Candolle detailed by himself, and we turn to them in the full expectation of finding ample justice done to his subject. Not that we may expect to learn that all, or indeed that very many physiological questions have been settled by him, be-

---

* Whilst preparing this article we have received Professor Lindley's " Introduction to Botany," in which the physiology of plants forms the subject of one book. The well-known proficiency of this eminent botanist will satisfy every one that he has here rendered an important service to this science.

yond the possibility of further cavil; on the contrary, the science is still so far in its infancy, that scarcely any of the most important laws and functions of vitality can be considered as fully understood. His work, however, is most valuable, in presenting us with a clear and explicit detail of the phenomena of vegetation, and a sufficient exposition of the various hypotheses by which different botanists have proposed to account for their existence. It is at once a compilation and a review of nearly every thing at present known on the subject. The work itself forms the second part of a complete " Course of Botany," which the author has for several years had it in his view to publish. The first volume of this course appeared in 1813, and a second edition of it in 1819, under the title of " Theorie Elementaire de la Botanique." This was succeeded by two volumes, entitled, " Organographie Végétale," in 1827. These three volumes completed the first part of the " Course." The present three, on " Physiology," constitute the whole of the second part; and the author proposes to publish hereafter a third part, containing " Botanical Geography," and other departments not yet discussed.

All researches that are undertaken in the several departments into which the study of nature is divided, may be classed under one or other of two general heads. They are either such as are made for the purpose of ascertaining the " sensible qualities" of bodies—as the materials of which they are composed, and their structure, whether internal or external; or else they are made with a view of discovering and estimating the laws which regulate the " various forces" acting upon, or by means of, the materials of which those bodies are composed—such as alter their condition, or produce in them various kinds and degrees of motion. We may say, therefore, that every branch in the vast study of nature has its " descriptive" and its " dynamical" department. The dynamical department of botany is denominated " Vegetable Physiology," whose immediate object is to determine the precise influence which the mysterious principle called " life" has upon the vegetable structure, under whatever conditions it may be placed. In his former treatises, our author had described the organs of plants, and their anatomy; in other words, he had shown us the construction of the machinery by means of which life is enabled to act, and to produce its effects; and in the present volumes he proceeds to show us this machinery in action. As, however, it seldom, if ever, happens that any single phenomenon in vegetation can be directly ascribed to an effort of " life" alone, but must also be considered partially dependent upon the simultaneous action of two other forces, " affinity" and " attraction;" it is a problem of no small difficulty to determine, what portion of the effect ought to be ascribed to each of these three forces independently of the other two. In all cases where a single force only is known to operate, its laws may be ascertained with precision, from the direct results which it produces; but when two combined forces produce a result, it is necessary that we should first ascertain the effect that would be produced by one of them when acting alone, before we can hope to appreciate that which is due to the action of the other. Now, the laws of " attraction" have been ascertained from the examination of the phenomena of the heavenly bodies, whose motions depend upon the immediate action of this force only. The laws of " affinity" are not yet ascertained; and, consequently, we cannot expect to determine, with perfect certainty, what those other laws may be which regulate the circumstances under which " life" is exhibited by the vegetable kingdom. When the physiologist would search for additional data for conducting his inquiry, and turns to the vital phenomena presented by animals, he finds his difficulties to be further increased by the presence of a " sentient" principle, which is in them superadded to the three forces above mentioned. Since, however, it would be impossible to await the final result of future researches into the nature and laws of " affinity," before we would proceed in our investigations of those which belong to " life" itself, we must endeavour, as well as we can, to carry on the examination of each of these forces simultaneously; being cautiously guarded not to ascribe to any effort of vitality what is actually due to the action of either of the two other forces; nor, on the other hand, be tempted to consider " life" itself as resulting merely from their joint operation upon a previously organized body. Every fact, then, connected with the existence of life in plants, should be carefully weighed in a double point of view: first, to see whether it has resulted from the action of one or more forces; and, secondly, how far each of these forces may have been modified by the peculiar structure of the body acted on. For the tissue itself of which the vegetable structure is composed possesses certain peculiar properties, and these must be first determined, lest they should be mistaken for vital properties. The physiologist, then, ought first to ascertain, as clearly as the case will permit, what portion of the result must be ascribed to the effects of attraction and affinity; what must be allowed to be the peculiar properties of the tissue itself; and, having abstracted these, he will arrive at what must be due to the specific action of the vital force.

In comparing vegetable with animal physiology, we find a striking analogy between many of the facts presented in each kingdom of organized nature; and since some departments of inquiry are more advanced in the one than in the other, the two studies may mutually assist each other in arriving at the solution of many questions of general physiology.

There are three properties more peculiarly distinguishable in the tissue of which the elementary organs of plants are composed. These are " extensibility," " elasticity," and " hygroscopicity." These it possesses whether in the dead or living plant, and therefore so much of every phenomenon as may be explained by their action should be ascribed to the inherent properties of this tissue, and not confounded with the functions of " life" itself. The cellular tissue of plants is enabled to accommodate itself to the development and growth of any organ, by its property of " extensibility," up to a certain point, beyond which it becomes ruptured, and must then be considered as dead matter. Upon its " elasticity" depends the action by which certain organs are maintained in particular positions, or by which they constantly return to such positions when any counteracting force is removed. Thus, in the common nettle, the stamens are curved forward in the early state of the flower, when they are held together by the anthers; but afterwards, as the filaments elongate, their elasticity alone is quite sufficient to dissolve the union of the anthers, and the stamens then fly back with violence and with a shock sufficient to cause the anthers to discharge their pollen. This particular movement therefore, and some others of a like kind, must not be ascribed to any specific vital action, but merely to the elasticity of the vegetable tissue. The " hygroscopic" action of the vegetable tissue is very considerable, and indeed constitutes its most important property. It ought not, perhaps, to be considered as any other than a peculiar case of " capillarity," where the pores which perforate the tissue are too minute to be capable of detection under the very highest powers of the microscope. It is most eminently conspicuous in the purest states of the tissue; as in the newly formed spongioles at the extremities of the root, and in the wood of the stem. We suspect, however, that De Candolle has in this latter instance confounded the action of the capillarity of the whole mass, originating in the presence of the intercellular passages, with the hygroscopicity of the tissue itself, and we can by no means assent to his explanation of gum and other matters being dislodged from the bark by an expansion of the wood, swollen by the hygroscopic action of the tissue composing it. Gums and resins, when secreted in superabundance, must necessarily be protruded externally, from the want of sufficient internal reservoirs appropriated for their reception.

After having noticed these three properties of the vegetable tissue, our author proceeds to examine three vital properties admitted in the animal kingdom, and which have sometimes been supposed to exist also in the vegetable. These are "excitability," "irritability," and "sensibility." "Excitability" is that property by which the "cellular tissue," forming the basis of the nutritive organs, manifests its susceptibility of certain impressions during the lifetime of the animal. "Irritability" is the property which causes the "muscular fibre" to contract upon the application of any stimulus. "Sensibility" is that property by which the "nervous matter" carries the knowledge of external impressions to the mind, or conveys the decrees of the will to the several organs of the body. Whether each of these three properties of animal life are also to be found in the living vegetable, admits of considerable doubt, and the difficulty of determining the question is principally owing to the great homogeneity of the vegetable tissue. The presence, indeed, of an "excitability" in the living principle of vegetation seems to be sufficiently clear: the cellular tissue resists destruction, and also effects certain chemical changes during the lifetime of a plant, which it is unable to accomplish after its death; and there are numerous phenomena which are totally inexplicable without admitting its presence. The boundary, again, between the animal and vegetable kingdoms is scarcely definable; and it seems impossible to allow this property to exist in those animals which seem to approach the nearest to the conditions of vegetable life, and yet to deny its presence in vegetables themselves. It is a property, however, which is much less active, though it is far more permanent, in vegetables than in animals. The existence of the vital property of "irritability" has often been suspected, though never clearly established, in the vegetable kingdom. In animals it resides in the muscular system, to which there is nothing analogous in the vegetable structure; and De Candolle considers all those phenomena which are usually cited as examples of an irritability in vegetables, to depend entirely upon the principle of excitability, modified in a few rare cases by certain remarkable peculiarities of organization, which are not yet well understood. But if "irritability" be not allowed to be a property of vegetable life, it seems still less likely that "sensibility" can. And, notwithstanding several plausible reasons for extending this vital property beyond the limits of the animal kingdom, it certainly does not appear to us that any facts which have been hitherto detailed can be brought forward with sufficient force to justify us in doing so, and we perfectly coincide with our author in believing, that so far as the present state of our information extends, "we can scarcely admit, without fresh proof, the existence of 'sensibility'

in plants, and that we consider them endowed with an ' excitability' only; that is to say, with a life analogous to that which
resides in such parts of the animal frame as are insensible."—
vol. i. p. 34. From these remarks, then, it follows that we consider the ultimate object of Vegetable Physiology to consist in
ascertaining the precise effects which the only known property of
vegetable life, its " excitability," is capable of producing.

The two great functions of life, to which all minor efforts are
subordinate, are " Nutrition" and " Reproduction;" and before any
inquiry can be instituted into the various causes which influence
and direct these functions, it is necessary that we should examine
the character of those elementary organs which are the seat of
the vital " excitability," and then, if possible, to ascertain the
manner in which these organs act. Now the internal structure
of vegetables is exceedingly simple, being entirely made up of
a congeries of cells and ducts, alike composed of a delicate and
apparently unorganized membrane. Most authors have considered
the ducts or tubes to be the chief seat of vitality, but De Candolle,
and, we think, most justly, ascribes to the cells the office of carrying
on the chief functions of life. Many plants are known to be
composed entirely of cellular tissue, and, therefore, in these at
least, we must allow that the vital force resides there, and consequently that all the more general functions of life are most
probably carried on by this tissue in all other plants also. But
although we agree with our author in considering the cellular
tissue to be the seat of the "excitability," we cannot assent to his
explanation of the mode in which he supposes it to act. His
idea is, that every cell is endowed with the property of succes-
sively dilating and contracting itself; the effect of which would
be to diminish and increase alternately the size of the intercellular
passages. This, he fancies, would cause a progression of whatever fluid may be introduced between the tissue of the plant.
But there are serious objections to this view, even if it were
clearly proved that such intercellular passages did always exist,
and were filled with fluid. We cannot understand how any close
cell or vesicle, containing an incompressible fluid, can be capable
either of contracting or dilating itself. The analogous examples
in the animal kingdom, which he produces in support of his
hypothesis, are not at all to the purpose. In these cases the con-
traction of the vessel is accompanied by an expulsion of the fluid
which it contains, and its dilatation by the introduction of the
ambient medium, whether liquid or gaseous; but in the kind of
peristaltic motion which he would attribute to each of the vesicles
composing the cellular tissue, no expulsion of the contained fluid
is supposed to take place; and as this fluid is also incompressible,

neither can we imagine how any dilatation of the cells can arise. We consider his attempt to explain the progression of the sap by the action of these vesicles, as the weakest part of the whole work. To say the least, it is purely hypothetical, and we also think it contradictory to known physical principles. On a subject of such extreme difficulty we ought not to speak very positively, but it does appear to us that M. De Candolle has treated somewhat too slightingly the recently discovered property of Endosmose, and has shown too great an attachment to an early theory of his own, which perhaps served very well as a conjecture at the time it was first proposed, but which he ought to have given up after a cause had been discovered, which is apparently sufficient to account for the general diffusion and progression of the fluids introduced into the vegetable tissue. We shall presently revert to this subject when we come to consider the progression of the sap. In whatever way, however, the cellular tissue may be supposed to perform its functions, we believe it to be the chief instrument through which the only vital power allowed to plants produces its effects. Light and heat appear to be the two great causes by which its energies are stimulated; and to these we must probably add electricity, and the more remote excitations induced by various chemical and physical actions.

Nutrition and Reproduction have been called " organic, or vegetative" functions, from belonging equally to the animal and to the vegetable kingdom. Notwithstanding the great differences observable in the external forms of organized bodies, the function of nutrition possesses certain general traits of resemblance in both kingdoms. In both, its operations may conveniently be separated into seven periods, and we shall attempt a brief sketch of the phenomena exhibited during each of these.

First period of nutrition.—This comprises the introduction of the food into the plant. As vegetables do not possess the power of locomotion, it is necessary that their food should be so generally diffused in nature, that they shall run no risk of perishing from their inability to search for it. Now it is a fundamental principle in vegetable physiology, that every thing capable of being imbibed into the tissue must previously be brought to a state of solution; but water is present every where in the earth, and in the atmosphere, and the material which constitutes the chief food of plants is carbonic acid, which is almost constantly to be found dissolved in all water. The root is the only true absorbing organ of this nourishment; for although, under certain circumstances, the leaf and some other organs may be made to absorb moisture, these are not to be considered as the organs originally destined for the introduction of food, any more than the pores in the skin of animals, which possess a like property, may be considered as their mouths. Neither is it by the whole surface of the root that this absorption takes place, but only through the " spongioles," which are small expansions of cellular tissue situated at their extremities. It is not clearly ascertained, whether the force which regulates the absorption of the spongioles is wholly vital, or whether it is the result of a vital action, in combination with the hygroscopic property of the cellular tissue. De Candolle had formerly attributed this action of the spongioles to their hygroscopicity alone; but he is now disposed to consider it dependant, to some extent at least, on the vital force. It does not, however, appear, that because there is a cessation of this absorption in the dead plant, and a continuance of it only in the living one, that we must therefore conclude it to be the result of a vital action in the spongioles themselves; for if it were really due to their hygroscopic property only, still the immediate removal of the absorbed fluid, by the organs which cause its progression through the plant, would continually renew the conditions necessary to secure a momentary repetition of its action. This absorption also has more of the aspect of a mechanical than of a vital operation, from the circumstance of all plants being equally indifferent as to the quality of the solutions which they imbibe, the quantity being regulated chiefly by the state of liquidity in which they occur: a more liquid solution of some substances, deleterious to the health of the individual, being more readily imbibed by it than others which are more viscous, though composed entirely of materials which are highly nutritious.

From the great uniformity in the means employed for absorbing them, and the general similarity of the matters absorbed, arises a great resemblance between the nutritive apparatus of all vegetables; which makes these organs ill adapted to the purposes of classification, and compels us to search among the reproductive organs for the characteristics necessary to establish a scientific arrangement of plants.

The usual aliment of plants we have stated to be water, containing carbonic acid in solution, or, we may add, at least containing some proportion of animal or vegetable matter capable of being converted into carbonic acid. But, besides this, air and various salts and other matters are absorbed in solution. Where, however, more substances have been found in the ashes of plants. than were supposed to have composed the materials by which they had been nourished, we must not conclude that the plants have created these substances, as some have imagined, but must consider them to have been extracted little by little, from some medium in which they really existed, though in such minute quantity as to be incapable of being detected by chemical tests. We may easily allow that plants surpass us in their powers of abstracting the minutest portions of any material disseminated through a given menstruum.

Second period of nutrition.—The water introduced by the absorption of the spongioles is called " sap, or lymph." It is then conveyed directly to the leaves, without sustaining any appreciable change in its progress, otherwise than by mixing with the vegetable juices it meets with in its course. The particular route which the ascending sap takes has often been a matter of discussion and dispute; but it has been clearly ascertained, by repeated experiments, that it ascends along that portion of the cellular tissue which constitutes the woody fibre, and not through the vascular tissue, which is intended primarily for the conveyance of air, though its tubes may occasionally be found filled with fluid. With respect, however, to the mode in which the sap is conducted along this cellular tissue, there is still much uncertainty. De Candolle favours the hypothesis of its passing along the intercellular passages, as we before mentioned, by means of the successive contractions and dilatations of the cells themselves. But we decidedly consider the newly-established principle of endosmose to afford a much better prospect of accounting for the fact. Endosmose is that property of all membranaceous substances, by which two opposite currents are established through their texture, whenever two fluids of unequal densities are placed on opposite sides of them. That the vital force must also be employed in securing any lengthened continuance of this action is evident; for in every apparatus constructed for the purposes of experiment, the effect must cease as soon as the two fluids have acquired the same density. Now, without attempting to account for the manner in which the healthy condition of the membrane is secured, we may easily imagine the constant development of fresh vesicles, and the continued secretion of fresh materials, to be sufficient to maintain the conditions necessary for the establishment of a perpetual endosmose during the lifetime of the plant, without considering this property itself to be directly dependant on the vital force. This property also explains the prodigious force with which the sap rises in certain seasons of the year; a force sufficient, as Hales determined, to support a pressure equal to two atmospheres and a half, and five times that by which the blood is propelled in the crural artery of the horse. The ascent of the sap is the result of a compound action, partly depending on a force " a tergo" propelling it forward, and partly on a force attracting it towards the foliaceous parts of the leaf; each, however, resulting from the endosmose carried on by all parts of the cellular texture.

Third period of nutrition.—When the sap has arrived at the leaves, and at the other green parts on the surface of the plant, a considerable portion of its aqueous particles is transpired. A cabbage, for example, transpires from a given surface seventeen times as much as a man by his insensible perspiration. A small portion, indeed, of this effect must be ascribed to the process of evaporation; but this is comparatively very trifling, and the greater part must be attributed to the action of a vital function. This is so decidedly remarkable in the vegetable kingdom, that De Candolle proposes for it the name of " exhalation," in order to distinguish it from the less conspicuous effects of the insensible perspiration of animals. It is manifestly produced by the instrumentality of the stomata, or glandular pores, seated on the green parts of plants, and which are more especially abundant on the under surface of the leaf. Heat exerts a trifling influence in producing an increase of exhalation, but light is the chief stimulant which determines its extent. Plants do not exhale moisture in the dark, and as they still continue to absorb a little, they soon become dropsical. The fluid exhaled is nearly pure water, and consequently the sap must become considerably altered by this circumstance alone, as the materials introduced in solution will now bear a higher proportion to the whole quantity of water retained in the plant. This great exhalation of the superabundant fluid may be considered analogous to the combined effects produced both by the insensible perspiration and excrementitial rejections of animals.

Fourth period of nutrition.—We have now arrived at the complicated phenomenon of vegetable " respiration," the most important of all the processes which together constitute the function of nutrition. One circumstance in this process is strictly in accordance with what takes place in the respiration of animals the presence of oxygen being equally essential to the life of the individuals of each kingdom, though the ultimate results are diametrically opposite in each. In animals, the oxygen inhaled unites with the superfluous carbon in the blood, and the carbonic acid thus formed is exhaled into the atmosphere. In plants, a similar effect takes place by night, when the leaves and other green parts inhale the atmosphere, whose oxygen unites with the carbonaceous matters dissolved in the sap; but the carbonic acid thus generated is, for the most part, retained in solution within the plant, and not exhaled again. All the coloured parts of plants perform this function as well by day as by night; but the green parts always decompose carbonic acid by day, from whatever source they may be able to derive it; and the result of this decomposition is to fix the whole of the carbon, and a small portion of the oxygen, in the substance of the plant, and to exhale the rest of the oxygen into the air. The chief supply of carbonic acid provided for this purpose is that which is introduced by the roots in a state of solution; but the small quantity universally present in the atmosphere is also inhaled and decomposed by the leaves. That which is formed within the plant itself is either retained in solution, or it is exhaled and disseminated in the atmosphere, whence it may re-enter the plant by one or other of the two methods just specified. Although the decomposition of carbonic acid is always proceeding during the day, yet it is never so rapid, nor so decidedly appreciable, as when the plant is exposed to the direct rays of the sun. It is independent of the presence of the stomata, which we have already described as being the true exhaling organs of the green parts; and it is certainly effected by means of the cellular texture of these same parts, whose green tint must be ascribed to the result of this very action. There can be no question that the " decomposition" of the carbonic acid is the direct operation of a vital function; but the " formation" of this gas in the coloured parts of plants, at all times, and in the green parts by night, appears to be the result of a chemical action similar to the ordinary process of decomposition in all dead organized matter. If a plant be exposed to the light in an atmosphere deprived of oxygen, it soon dies, unless (which is very remarkable) it be enabled first to form a little oxygen by decomposing some portion of the carbonic acid within it, and thus to impregnate the atmosphere with a sufficient quantity to enable it to act as a further resource and constant stimulus for the formation and decomposition of fresh supplies of carbonic acid.

The result of all these compositions and decompositions of carbonic acid in the living plant, is the fixation of the whole of that carbon which is found in the entire mass of vegetation on the surface of the earth. Thus the atmosphere is continually being purified of every fresh addition of this material with which combustion, respiration, and putrefaction are perpetually adulterating it. For though all living plants do themselves generate carbonic acid in the way which we have specified, they also decompose much more than they form, and this excess is considered to be sufficient to counterbalance the quantity introduced into the atmosphere by other causes. In this single function, then, of vegetable life, we see an efficient yet simple means of restoring to the atmosphere that proportion of oxygen which is necessary to the health and existence of organized beings.

Thus, the consideration of the humble functions of vegetable life may serve to elevate our thoughts in wonder at that universal order which prevails throughout the works of the great Creator and Preserver of all things.

Fifth period of nutrition.—This period comprises the return of the newly elaborated sap into the system. The course which it pursues in its altered condition is, for the most part, down the innermost layers of the bark; as several very satisfactory experiments have decided. In the process of " ringing," which consists in stripping a branch of a circular rim of bark, the descending sap is stopped, and a swelling is formed at the superior edge of the ring. The alburnum, or soft wood immediately below the bark, will also, if properly protected, serve to convey a portion of the descending sap, which is then further elaborated, and the alburnum hardens to solid wood. It has been a subject of considerable controversy, whether the new layers of wood and bark have resulted from the development of the old tissues nourished by the descending sap, or whether they have been generated partly in this way, and are partly formed of fibres descending from the buds which are seated on all parts of the stems and branches. Perhaps the question ought not to be considered as finally settled; but certainly we agree with our author in considering the latter opinion, of the descent of the fibres, as one which is little supported by facts, or by analogy, and that it rests for the present entirely upon vague conjecture and hypothetical reasoning. The elaborated sap descends to the roots, and causes their further development, whilst a portion of it is intercepted by the rising sap by which it is conveyed to the buds on the stem. There is, in fact, no true circulation in plants; but a portion of the elaborated sap, in its descent towards the roots, becomes mixed with the ascending sap, and is thus conveyed to all parts of the system.

Sixth period of nutrition.—In detailing the fourth period, we accounted for one alteration in the condition of the sap, by the circumstance of a considerable exhalation of its aqueous particles taking place, and for another, by its having received an addition of carbon from the decomposition of carbonic acid in the green parts. The nutritious material thus formed is essentially composed of carbon, and the two elementary ingredients of water, viz., oxygen and hydrogen. There are, however, several vegetable products, differing materially in their sensible qualities, which are composed of these three elementary substances only, and it is a task of some delicacy to select that particular one from among them which may most reasonably be considered as the universal pabulum prepared for the nourishment of the different vegetable tissues. Our author considers " gum" to be the simplest combination of the three elementary ingredients mentioned ; and argues, from its universal prevalence, that it must be the true nutritious principle of vegetables. There are some other substances nearly allied to gum in their chemical composition, which appear to be slight modifications of it, effected in some after process, by the secreting powers of the cellular tissue. The preparation of these fresh substances constitutes our sixth period. They seem to serve some purpose connected with the nutrition of the plant; but what this may be, it is impossible, in the present state of our knowledge, to decide. Three of these substances are fecula, sugar, and lignine, each differing very little from gum in their chemical composition. If we consider the solution of gum, so constantly found in the sap of plants, as analogous to blood, the formation of these other materials may then be likened to certain local secretions in the animal kingdom. Each grain of " fecula" appears to be a reservoir of gum incased in an insoluble integument. It is diffused through various parts of the plant, serving as so many magazines of nutriment for the future development of its several organs. It bears a striking analogy to the fat of animals. " Sugar" bears a strong resemblance both to gum and fecula in its composition. It is found in a liquid state in the cells, and is probably intended to serve some purpose or other of nutrition. " Lignine" is insoluble in water, and is a secretion deposited in the cells, which compose the woody portions of the plant.

Seventh period of nutrition.—The three substances mentioned under the sixth period, appear to be destined to serve some purpose or other of nutrition, as well as that universal pabulum, " gum," of which they are only slight modifications. Besides these, there are many other substances which result from the specific action of distinct parts of the vegetable structure, and which bear a still closer analogy to the peculiar products secreted from the blood of animals by the action of particular glands. In animals, however, the glands destined for this purpose are very conspicuous, the ducts through which the secreted matter is conveyed being clearly defined, and the secretions themselves presented to us in an isolated form. But in plants, the glands are generally minute, their structure scarcely distinguishable, and many of their secretions so much blended with other materials, that it requires a chemical process to separate them.

The enumeration of these various substances occupies a considerable portion of the first volume ; and an attempt is made to classify them under a few general heads ; but as no light is thus thrown upon the function of secretion, we shall allude very briefly to these details. Every separate vesicle of the cellular tissue may be considered as a secreting organ, and some of them appear to elaborate peculiar compounds without assuming any of the ordinary characters of distinct glands. In other cases, a glandular structure is clearly distinguishable from the rest of the tissue. The matter elaborated is either destined to appear on the outside as an excretion, or it remains within the plant, but is so arranged in separate cavities as not to intermix with the nutritive juices. These all differ from those other secretions which we have considered as eminently nutritive, by having their oxygen and hydrogen in a different proportion from that in which they exist in water; and hence it seems probable that they result from a later and more complicated process than that which produces the nutritive secretions belonging to the former period. They all moreover act as poisons when imbibed by the roots; and this again shows us, that in the living subject they must necessarily be contained in specific cells provided for their reception, and that they can form no part in the process of nourishing and developing the plant. There are certain local secretions which can only be separated from the general admixtures in the sap by chemical processes; such are the various vegetable acids and alkalis, the origin of which is not at all understood. In the ashes of plants also we find variable quantities of different earths, metals, and salts, all of which have been introduced in solution with the water absorbed by the roots. A supposition which has been made, that some of these materials may be the direct produce of an effort of vegetable life, is wholly untenable, and their presence is clearly to be accounted for on the principle of their absorption in a state of solution. It is a more delicate question to determine, whether these materials ought to be considered as merely adventitious, and unnecessary to the health of the plant, or whether their presence is really beneficial to it. When received into the system they are conveyed by the sap to the leaves and surface of the stem, where they are always found to be deposited in greater abundance than in any other parts of the plant; which arises from the constant exhalation of the water in which they were dissolved taking place there. Hence, the annual fall of the leaf secures a constant discharge of these earthy matters from the plant, and a renewal of those organs takes place, which otherwise must ultimately have become choked by them.

Having completed the account of the various processes into which the great function of nutrition may be separated, some account is given of the progress made in the annual growth of a plant during each of the four seasons.

In " winter," the vital action remains nearly torpid. A continued but feeble absorption takes place at the root, sufficient merely to supply the slight degree of exhalation still carried on by some of the organs seated towards the surface. In " spring," the increased temperature is the great stimulant to the vital excitability, and the bark now begins to attract the sap towards it, and a fresh current proceeds from the roots. The formation of new radicles and spongioles adds increased vigour to this flow of the sap, and the immediate consequence is, the development of the buds. Independently, however, of these stimulating causes by which the vital energies of the plant are roused to action at the return of spring, it should seem that there is a special law of vitality, predisposing the plant to make these efforts after certain periodic intervals. Their success also depends, in a great measure, upon the conditions under which the plant has been placed during the previous year. During the " summer," the function of nutrition gradually diminishes in the intensity of its action. By the " autumn," the leaves have become choked by the earthy particles deposited in them; their fall commences, and the true sleep of the individual takes place. This, as is well known, is the best season for transplanting, when the juices are stationary, and the new fibrils have not yet been developed on the roots.

In explaining the mode in which the young branches and roots are developed, it is to be noticed that the former expand throughout their whole length, while the latter are increased by successive additions at their extremities only. During the process of development, something like periodic returns of intensity has been remarked, and it is stated that these efforts are twice accelerated and twice retarded every day; but the account wants confirmation.

There is another phenomenon, however, and that of a most curious description, which ought to be considered as the result of a specific vital action; though some observers have supposed it may be accounted for on physical principles. We allude to a constant rotation of the fluids contained in the cells and short tubes of some plants. This rotation is rendered apparent by the presence of little globules, or rather granules, of vegetable matter swimming in the fluid, and it may be seen very readily in all the species of *Chara*, under any ordinary microscope. It is found also in several other families of *Cellulares;* and has also been detected in the genera *Caulinia, Hydrocharis,* and *Valisneria,* among *Vasculares:* to which we may add an observation of Mr. Brown, who has noticed a distinct rotation of the fluid contained in the joints of the moniliform hairs which clothe the base of the filaments in *Tradescantia*. These details conclude the first volume, which closes the account of the all important function of nutrition.

II.—We now proceed to consider the second of the two chief functions of vitality—" Reproduction." The continuation of the species is secured in the vegetable kingdom by two distinct methods: the one by the " fructification" of the plant, by which its reproduction, strictly speaking, is effected ; and the other, by the " subdivision" of the plant, by which its multiplication is obtained without the intervention of any reproductive process. These two means of propagating plants should be separately examined, and then the conditions essential to secure success in each case should be compared, in order that our ideas of what is a " species," and what a " variety," may be fixed. We shall follow our author in his account of the function of " Reproduction," by separately considering the five periods into which he has divided the process of " fructification," viz. flowering, fecundation, maturation, dissemination, and germination of the seed.

First period of fructification.—This comprises the well known phenomenon of " flowering," in which one of the fundamental organs of the plant, we mean the leaf, by the operation of certain causes totally unknown, has ceased to obey the ordinary conditions of development essential for securing the nutrition of the individual, and has assumed a totally different character, in which it performs entirely new functions. As the predisposing causes which determine this phenomenon are unknown, we must be content to compare the actual period of a flower's expansion with the age of the plant, and the time of its duration ; without pretending to say why or when the nascent organs first obtained the character of a flowering bud.

The flowering of plants may be compared to the age of puberty in animals; and the laws which regulate its commencement, as well as its periodical returns in perennials, are sufficiently definite for each species, though liable to considerable modification under peculiar circumstances. An elevation of temperature, for instance, decidedly excites and accelerates the flowering propensity, whilst a superabundance of moisture counteracts and retards it. Hence it happens that when the fruits and legumes of a temperate climate are cultivated within the tropics, although the increased temperature might dispose them to flower more freely, yet the moister atmosphere of those climates predetermines their running to leaf. Whatever, on the other hand, tends to check the luxuriance of vegetation, predisposes the plant to show more flower; and hence the effects of our winters are undoubtedly beneficial in allowing the sap time to be prepared for this important function. In India, it is customary to uncover the roots of fruit-trees during the violent heats, which causes the leaves to fall, and thus the effects of winter are artificially produced upon the constitution of the plant. The periodic return of the flowering season is sufficiently constant in each species, to be considered as a general law of vegetation; for, although it is subject to modification from a variation in temperature, yet it is evidently dependant also on the idiosyncracy of the individual. De Candolle mentions two contiguous trees in a row of horse-chesnuts at Montpellier, one of which always flowered the earliest, and the other the latest, of all the trees in the row. It is important, in an economic point of view, to note this variety in the periodic times of flowering in different individuals of the same species, since cultivators, by continually taking advantage of it, and so propagating from the earliest and latest of the same crop, may gradually obtain plants which will yield a much longer succession of flowers or fruit than could otherwise be obtained.

Light, and not heat, appears to be the stimulating cause most effective in determining the particular hour of the day in which flowers expand. Some never expand if the weather is likely to prove rainy, though a sudden storm will occasionally surprise these natural hygrometers, cheated by a brilliant sunshine into a state of unsuspecting security.

In general, the flowers are developed with a regularity conformable to the rate of increase of the whole plant; but in some species, the vital energy appears to become excited in an unusual degree, as the period for flowering approaches. Thus, the American aloe (*Agave Americana*), which in the south of Europe continues vegetating for three or four years without flowering, and in our own country for forty or fifty, and which during this period has increased very slowly, suddenly exhibits a rapidity of development, when the flower is about to make its appearance, which is very remarkable. In three or four months its flower stalk acquires a length of fifteen or eighteen feet. A plant of the *Agave fœtida*, which had been in the Paris garden for nearly a century without showing any indications of flowering, and which had scarcely appeared to grow at all, in 1793 acquired a flower stalk twenty-two feet and a half in length, in the space of eighty-seven days, and was observed at one time to be growing at the rate of more than one foot per diem.

Second period of fructification.—When the flower has expanded, and its various organs have attained the requisite development, the important process of fecundation takes place, by which the fertility of the seed is to be secured. M. De Candolle prefaces his account of the fecundation of the ovule by a brief historical review of its discovery. It had been ascertained from the earliest antiquity that some preliminary act was in all probability required for impressing a vitality on the embryo contained in the seed, before it could become capable of a separate and independent existence. No certain information, however, had been acquired as to the precise nature of the process; and the few notions which prevailed were very vague, and for the most part erroneous. Herodotus confounds the phenomenon of the caprification of the fig, practised in the East, with the fecundation of the date, obtained by suspending bunches of the barren, or male flowers, over those trees which bear the young fruit. But it is now ascertained that the caprification of the fig is serviceable merely in accelerating the ripening of its fruit, owing to the puncture which it receives from the larva of a *Cynips*, a small insect abounding among the blossoms of the wild figs, which are suspended over the cultivated plants in the flowering season. We shall add nothing further to our account of this history of the final establishment of the sexual distinctions of plants, because the subject has been so often discussed, and is now so securely settled, and so generally known as to need no fuller notice. But before we proceed to examine those facts which have been most recently discovered concerning the mode in which the process of fertilizing the ovule is carried on, we must advert to the experiments of M. Lecoq, because De Candolle appears to consider them to be worthy of credit. They tend to prove that the female plants of certain diœcious annuals, as the hemp, are capable of ripening their seed without the necessity of any previous action of the pollen. But these rare exceptions to an otherwise general law militate nothing against the well established doctrine of the sexes in plants. They may show us the necessity of modifying our ideas on the subject: but they no more disprove the fact itself than the phenomenon which results from the impregnation of an *Aphis*, whose offspring are capable of producing young without fresh impregnations to the fifth generation, can persuade us of the non-existence of sexes in the animal kingdom. It is quite right that we should be mistrustful of maintaining the existence of too strict an analogy between the two kingdoms of organized beings; but there is no point of analogy between them that is more striking than this; and it would be far more advantageous to the progress of science, if certain modern objectors would direct their researches towards examining the process by which the fecundation of the ovule is effected, instead of wasting their time in framing doubts of its existence.

We find various provisions made for securing the success of the fecundating process. In some plants, one or other of the organs destined to this purpose are endowed with specific motions, by which they are brought nearer to each other at the time. In most others they are placed so advantageously, that there is no chance of failure, under ordinary circumstances. Various precautions are provided to enable them to avoid the deleterious effects of moisture, which might cause the pollen to explode, if brought into contact with it. The organs destined for securing the fertility of the seed are, as every one knows, the stamens and the pistil; the essential part of the former being the anther, which contains a fine dust named the pollen, and each grain of pollen is filled with exceedingly minute granules, called the fovilla. The pistil is crowned by the stigma, which is the organ destined to receive the grains of pollen on its surface; and then the process of fecundation consists in that influence, whatever it may be, which is exerted by the fovilla being carried through the pistil down to the nascent ovules seated within the ovarium. There is a material interruption in the chain of evidence hitherto brought forward, which would account for the manner in which the influence of the fovilla is conveyed to the ovule; and three hypotheses have been proposed, each of which is at present without sufficient proof to warrant its unqualified reception. Observers have long agreed that the grains of pollen, when placed in water, discharge their fovilla with violence through certain pores on their surface; but Amici first discovered, in 1823, that when they fall upon the stigma, and are left attached to it by means of a glutinous exudation which coats its surface, they do not explode; but that after the lapse of a few hours, they protrude from their surface one or more exceedingly delicate tubes, which insinuate themselves between the vesicles composing the stigmatic tissue, and subsequently extend to a considerable length down the style. From later observations of the same, and other observers, it has been shown that these tubes in some cases actually penetrate into the ovarium, and that they there appear to distribute themselves between and around the ovules. The fovilla enters these tubes; but whether its granules ever reach the ovules, as M. A. Brongniart supposes, or whether they only produce upon them a certain stimulating effect, as Mr. Brown imagines, are questions not yet decided. In researches of such extreme delicacy, no direct observations seem likely to bring us to a satisfactory solution of the difficulty, and it is only by arguments derived from analogy, and by a comparison of the whole of the phenomena, that we can expect to solve it. Before we dismiss the subject, we must refer to two remarkable papers alluded to by De Candolle in a note at p. 524, but which he did not receive in sufficient time to incorporate the discoveries made by their authors into the body of his work. These papers, the one by Mr. Brown, and the other by M. A. Brongniart, contain successful attempts to reduce the mode of fecundation in *Orchideæ* and *Asclepiadeæ*, under the laws previously ascertained to exist in other plants. The two Orders here alluded to, from the anomalous structure of their pollen, and several other peculiarities, had long puzzled botanists, and it was very generally supposed that their mode of impregnation was totally different from that of other plants. The labors of the two eminent botanists here mentioned have separately and satisfactorily shown us that this is not the case ; and thus our former notions on the subject have been confirmed in a very unexpected manner. There are some facts so very remarkable in the appearances detailed, that Mr. Brown, whom no one will ever accuse of broaching an extravagant or hasty opinion, but who is deservedly celebrated, beyond all other observers, for the extreme caution with which he conducts and publishes his researches, has here ventured to hazard an opinion, which we shall give in his own words.

"In conclusion, I venture to add, that in investigating the general problem of generation, additional light is perhaps more likely to be derived from a further minute and patient examination of the structure and action of the sexual organs in Asclepiadeæ and Orchideæ, than from that of any other department either of the vegetable or animal kingdom."

Although it is natural to suppose that the various floral envelopes, the bracteas, calyx, corolla, and nectaries, are in some way or other serviceable to the plant in the process of fecundation, yet their specific functions have not been ascertained. It seems, however, not to be improbable that they exist in some families without any specific use, and are intended only to mark that general symmetry which prevails throughout the different departments of the vegetable kingdom.

Third period of fecundation. As the gradual maturation of the fruit commences from the moment after the ovule has received the fecundating stimulus, it may be considered as analogous to the period of gestation in animals. The great majority of plants ripen their fruit within a year after the fecundation of the ovule has taken place; but many species of the pine tribe are more than a twelvemonth in perfecting their seeds. The cedar is the most remarkable example of this fact, which carries its cones for twenty-seven months after they have flowered, before their seeds are spontaneously shed from them.

So soon as the embryo appears, or rather from the instant that the ovule has received the fecundating influence of the fovilla, it becomes a centre of vital activity; and, like all parts of organized bodies that have become specially excited, it immediately attracts from the surrounding parts those juices which are required for its own nutrition. The development of the ovarium also takes place, and it now assumes the character of a pericarp; and those ovules which have received the fecundating influence enlarge to the condition of seeds, whilst the rest of them . decay, and are soon shrivelled up. The great regularity with which a specific number of the ovules always becomes abortive in the same species, is very remarkable; thus the ovarium of the oak contains six ovules, but the pericarp of the matured acorn contains one seed only; the ovarium of the horse-chesnut contains six ovules, but it rarely happens that more than one of them is perfected, the other five having been strangled in succession, whilst the pericarp was enlarging.

The pericarp of some fruits will enlarge and ripen without requiring the impregnation of the ovule to have taken place, whilst in others it will not increase unless this has been effected. The cultivated pine-apple is an example of a pericarp not only ripening, but greatly improving by the abortion of the ovules; for the wild variety, in which alone the seeds ever ripen, has its pericarps very little developed, and the fruit is proportionably dry. The bread-fruit cultivated in the Friendly Islands, is another example of a variety in which the seeds are abortive, and the pericarp in consequence proportionably increased. Other plants, as corn, produce no fruit unless the fecundation of the ovule has been secured. So long as the pericarp continues green, its action is precisely the same as that of the leaf; it exhales the superfluous moisture, and decomposes carbonic acid; but as it gradually ripens, its cellular texture begins to elaborate the sap into those various materials which form the peculiar qualities of different fruits. An increased temperature both accelerates and improves the ripening of fruits; and the puncture of insects stimulates the secreting function, and produces an earlier fall. It is this effect which, as we have already observed, follows the caprification of the fig practised in the Archipelago, and which enables the cultivator to procure two crops in one year, by its causing the spring buds to ripen so much earlier, that time is allowed for the summer buds to ripen also, which otherwise they would not be enabled to do.

The process of ringing, discovered in 1776, which consists in removing a narrow ring of bark from a branch about the time of flowering, checks the descent of the sap, accelerates the maturation, and often enlarges the size of the fruit. When the ring has been judiciously made, the bark unites again over the wound, and no apparent injury results from the operation. De Candolle considers the grounds upon which this practice has become unpopular as insufficient to condemn it, and proposes that fresh researches should be undertaken to show the precise extent to which it may safely be trusted.

The chemical alteration which takes place whilst the process of maturation is proceeding, consists principally in a diminution of water, and an increase of sugar, at the expense of the lignine and other matters.

Passing from the ripening of the seed covers to that of the seeds themselves, we must first consider what was the state of the ovules previously to their being fecundated, and then trace the changes which they gradually undergo till they have assumed the condition of ripened seeds. In its earliest state the ovule consists of a fleshy nucleus surrounded by one or two coverings, which are pierced by an opening at one end, which end is called the apex of the ovule. After the fecundation of the ovule, the embryo sooner or later makes its appearance within the nucleus and near its apex, and is then surrounded by a membranaceous coat called the embryonic-sack. The materials prepared within the ovule for the nutriment of the embryo are the "amnios," which immediately surrounds it, and the " spermoderm," by which the amnios itself is surrounded. As the embryo enlarges, these materials are more or less absorbed by it; but in some seeds, where there remains a considerable supply over and above what was required for its full development, this nutriment becomes the albumen; that substance which so remarkably and providentially supplies the wants of man in all the various kinds of grain which he cultivates for food; but which otherwise was destined to nourish the young plant during the early stages of its germination. When the seed is ripe, all the free water which it contained has been elaborated either into fecula, or into a fixed oil, and the completion of this process may be considered as the termination of its maturation; in which stage, all the nutritious materials previously stored in the stem for this purpose have been exhausted.

There are various contrivances provided for protecting the fruit from injury which we need not here allude to, but shall proceed to the

Fourth period of fructification.—It is not sufficient that the seed should be perfected and protected, and the soil prepared in which it is to germinate: there are other conditions necessary to secure the continuance of the species; among which are to be ranked the various modes appointed for the dispersion and dissemination of the seed, which will form the fourth period in our account of the progress of the fructification of plants. A prosecution of the inquiry into the various laws which regulate the co-existence of certain forms and organs, with the relations which they bear to each other in securing the continuation of the species, is connected with the loftiest questions of metaphysics, as well as with the most abstruse points in natural history. We cannot here pretend to describe all the various means by which the dispersion of the seed is secured, but we may notice very briefly the three divisions into which our author throws fruits in general, as connected with their mode of dispersion and dissemination. In his *first* division, the pericarp invests the seed so closely as not readily to be separable from it; as in the case of corn and many other plants, whose seed-covers are sown together with the seed itself. Seeds of this kind are frequently furnished with chaffy, membranaceous, or feathery appendages, by which they are wafted by the wind to considerable distances. His *second* division consists of those fleshy and succulent fruits whose covering rots off, and appears to be of no use in promoting germination, and which must be removed if we wish to preserve the seed from rot. These fruits are in a less favourable condition for being dispersed than any others; though there is one obvious mode by which this may be effected, when, after they have been swallowed by animals, their kernels are voided containing the seed in an undigested state. His *third* division includes those fruits in which the seeds are contained in dry capsules, many of which open in a way that best admits of their being gradually scattered. Some capsules burst in dry and others in moist weather, according as the nature of the plant requires a dry or moist condition for its germination. A remarkable instance of the latter description occurs in the plant vulgarly known by the name of the Rose of Jericho (*Anastatica Hierochuntina*).

"This little plant grows in the most parched deserts. By the time it dies, owing to the great drought, its tissue has become almost woody; its branches fold over each other till the whole mass assumes the form of a ball; its seed vessels have their valves tightly shut, and the plant remains adhering to the ground by a solitary branchless root. The wind, which always acts powerfully along the surface of a sandy plain, uproots this dry ball, and rolls it along. If it now chance to meet with a splash of water whilst performing its constrained but necessary journey, it speedily imbibes the moisture, which causes the branches to unfold, and the pericarps to burst; and the seeds, which could not have germinated if they had fallen on the dry ground, now sow themselves naturally in a moist soil, where they are able to grow, and where the young plant may support itself."—p. 613.

Whatever wonder the superstitious accounts recorded of this plant may have excited in the ignorant, it is assuredly surpassed in our thoughts by a steady contemplation of that wisdom which has so marvellously provided that a humble weed should thus, by means so simple, secure the preservation of its species amidst the many chances of destruction by which it is surrounded.

Some seeds, if thoroughly matured and placed under such conditions as may secure them from the united influence of water, heat, and oxygen, the three requisites in germination, seem capable of preserving their vitality for an almost indefinite length of time. There are numerous instances recorded of seeds, which must have lain dormant for centuries, having immediately vegetated when the ground has been trenched for the first time within the records of man. It is stated on the authority of Gerardin, that French-beans, which had been in the herbarium of Tournefort for at least a century, have been caused to germinate; and De Candolle informs us that a sack of the seeds of the Sensitive-plant, brought to Paris sixty years ago, still furnishes the annual supply required for raising young plants. The preservation of seed in a sound condition is of the highest importance in an economical point of view, and can be brought to perfection only by studying the laws by which nature provides for their dissemination and germination. The main object is to secure them completely from moisture, and also as much as possible from the influence of oxygen. Various methods have been proposed for preserving them on a great scale, the best of which consists in enclosing them in large wooden cylinders hermetically sealed, or else in sacks impervious to moisture and the influence of the atmosphere.

Fifth and last period of fructification.—Few seeds, comparatively speaking, escape the various dangers to which they are subject, either from being devoured by animals, or from falling upon a soil unfit to receive them, or from perishing by exposure to the inclemencies of the weather. In such, however, as escape these dangers, the process of germination commences as soon as they are supplied with the three requisites of water, oxygen, and heat, in a proper proportion. This process consists in the revival of the embryo from its state of torpidity, when it bursts its integuments, and for a short time exists only upon the food previously laid up for it in the seed, until at length the development of its nutritive organs is completed, when germination is considered to cease, and the function of nutrition begins to be carried on in the ordinary way. This period may be considered analogous to that of suckling in quadrupeds, or more strictly to that of incubation in birds. After an absorption of moisture has taken place, at least equal to the weight of the seed, the descent of the radicle and ascent of the plumule begin, which constitute the character-istic phenomena of germination; and the exhaustion of the coty-ledons, or the consumption of the albumen, if there be any, at length completes the process. Although there are other circum-stances usually attending this phenomenon, they do not appear to be among the conditions absolutely essential to its success. The soil, for instance, is of material assistance in supporting the plant, in regulating the quantity of water with which it is sup-plied, and in furnishing the materials for its future growth; but still the want of a soil is not fatal to germination, though without one we cannot expect a successful issue during the future pro-gress of the plant. The action of light, again, has either no effect on germination, or else it is actually injurious to it: for light promotes the decomposition of carbonic acid, and the fixa-tion of carbon, whereas the process of germination requires the presence of oxygen, in order that it may become united with the superfluous carbon already in the seed and so carry it off. Water appears to be absolutely requisite for diluting the nutriment con-tained in the cotyledons, or in the albumen; and heat probably acts only as a stimulus. When the certainty of germination is once secured, the radicle is the first part of the embryo which is developed.

We stated that there was a second means of propagating plants, by subdividing them; and we shall now enter upon an examination of this process, and shall compare the results with those obtained from the mode of reproduction just described.

It has been ascertained that the outer coat of the seed, the cotyledons, and the albumen, may each be removed from the embryo when put into a state of germination, without absolutely destroying its life, though the plant can never afterwards recover the shock, but always remains stunted and dwarfish. If the plumule be removed, a further development of the radicle will still take place, and if the radicle be removed, the plumule still continues to increase; at the same time also an effort is made to restore the mutilated part. There is of course a limit beyond which these mutilations cannot be carried without destroying the life of the plant; but these facts tend to show us that life is actually present in every part, and that it makes an effort to supply the loss of any organ that is removed, by the formation of a new one. It is upon the success of this effort that the property of "multiplication by division" depends: where those organs which are wanting in the detached part become renewed by the appro-priation of part of the nutriment originally stored in this part for its further development. The power of multiplying by division, exists, indeed, in a small number of individuals of the animal kingdom, and forms one of the most remarkable characteristics of vegetables, among which it appears to be almost universally possible. De Candolle proposes to restrict the name of "tubercle" to every kind of gem or bud which is capable of being developed into a perfect plant when detached from the parent stock. The peculiarity of the tubercle consists in its first developing its ascending organs, by which it differs remarkably from the seed, in which the descending organs are always first protruded.

The reproduction of the descending organs takes place whenever a portion of the stem is exposed to moisture, and may proceed also from other organs, as the leaf. Although there are evidently very considerable differences attending these two modes by which a plant may be propagated, by seed or by division, yet in the formation of the embryo and of the tubercle upon the same individual, these bodies evidently exert a mutual influence on each other. The law which regulates this influence is not known, but it is most probable that in all plants which produce both seeds and tubercles, the one sort fails in proportion as the other succeeds; there seems, therefore, to be considerable plausibility in the notion that each of them is only a peculiar state of some common germ of vitality, modified by circumstances, of which we are, and perhaps ever shall remain, utterly ignorant.

If the laws which regulate the reproduction of flowering plants are to be considered as undetermined, we must allow that scarcely any thing is known of those which provide for the reproduction of the cryptogamic tribes. Many species among them possess two modes of reproduction, the one by the formation of sporules, and the other by the subdivision of their parts. There appears to be no solid foundation for the old notion, now, indeed, scarcely entertained by any one, that some species originate from spontaneous generation; for though the assertion, "*omne vivum ab ovo*," may, perhaps, be justly considered too hazardous, yet every well-attested fact tends to show us that the appearance of a fresh individual on the surface of the earth must be the result of one or other of the two modes of reproduction here described.

The study of "species" forms the basis of every scientific classification of plants, and one inquiry of the vegetable physiologist should be directed to ascertain the limits within which the characters of a given species may vary. Botanists have hitherto sadly neglected the only sure means of bringing this question to a satisfactory issue, namely, the test of careful experiment. Various theories have been framed by different observers, some of whom suppose with De Candolle that originally there were a certain number of primitive types, or forms of species, created, from whence all subordinate varieties have resulted; and although, as in hybrids, a form intermediate between two of these original types may arise, yet they imagine that no very considerable or essential deviation from them has ever been subsequently introduced; and also, that no entirely new type has ever originated from them. Others again, with Linnæus, have not only imagined that many new forms have resulted from a combination of different species, but that, by crossing the species of different families, even new genera may have arisen. All hypothesis, however, on this subject, remains hitherto unsubstantiated by any thing like the satisfactory evidence of experimental proof, which can be considered the only legitimate basis of philosophical speculation. In our present state of knowledge, the various modifications to which a species is subject, are severally called its varieties, races, variations, deformations, monstrosities, and hybrids. Some of these kinds of modification originate from the influence of external causes alone, others are the result of certain peculiarities and combinations connected with the fecundating process; and, as these latter belong strictly to the inherent qualities of organized bodies, they are consequently more difficult to comprehend, and impossible to be accounted for. According to the notions by which botanists are at present guided in their classifications of the different individuals of the vegetable kingdom, we may often find two distinct species, apparently differing less from each other than two varieties of one and the same species; but whether such notions are correct must depend entirely upon the accuracy of our idea of what a species is. De Candolle considers a species as embracing every possible variety of form which may be produced from seed prepared by a single individual, or by a couple of individuals. But this definition includes all hybrids, and would therefore compel us either to consider the parents of a hybrid to be of the same species, or the hybrid itself to be a newly created species. He is, moreover, disposed to consider all permanent varieties, or true races of plants, as hybrids more or less removed from the parent types, and which have become capable of reproducing their kind by seed. As a conventional term, and until the subject is better understood, the ordinary idea attached to this word had better remain; which supposes a single species to embrace all varieties which sufficiently resemble each other, to allow of our considering them to be not improbably the offspring of a single individual, or (in the diœcious species) of a pair of individuals of the same kind. Hybrids we would still class apart from true species, until the laws which regulate their formation shall become better understood. The permanence of true species seems to be strongly attested by the fact, that any of those modifications in form which have resulted from the operation of external causes, may often be removed, by subjecting the individual to the effects of other causes of an opposite description. As when plants, for instance, which have assumed one character whilst growing in a moist situation, are made to assume another when they are transplanted into dry ground. The production of hybrids, also, and of such varieties as constitute distinct races, never introduces any entirely new form, but merely modifies those which are already in existence. The permanent character of particular forms has also been fully established by authentic records in the history of the last three thousand years; a fact which must outweigh any vague conjectures or unsubstantiated theories to the contrary. More than eighty plants have been recognized on the monuments of Egypt as being identical with such as are still in existence, and the fragments of twenty species at least have been actually found among their mummies. Our ignorance of the origin of species is no argument for concluding them to have been produced by the agency of external causes, and it is far better for us to confess these facts to be beyond the limits of our knowledge, and to confine our attention to what is clear and comprehensible. Hybrids are manifestly analogous to mules in the animal kingdom, and their rarity in a wild state is nearly equally great in each case, although they may readily enough be procured by art. As the smallest possible quantity of pollen is sufficient to fertilize the ovules, and as these, when once impregnated, are rendered incapable of receiving any fresh influence, it could not often happen that the stigma would receive the pollen of another plant, with which it might be capable of hybridizing, before some portion of its own must have fallen upon it.

Besides the several peculiar phenomena which belong to the separate functions of nutrition and reproduction, there are others which may be considered as equally allied to either of them. The consideration of these forms the subject of a separate book in our author's second volume.

A marked law of symmetry regulates the conditions under which the vegetable structure is presented to us, in such plants as are closely allied in natural affinity, however much they may differ in certain individual peculiarities; these peculiarities always depending upon some modification in the mode of development in certain organs, or upon the partial or entire suppression of them in the one and not in the other species. Repeated examples have shown us, that certain organs may sometimes be accidentally developed in plants in which they are generally absent, or else may disappear in some individuals of a species where they are usually present. It is by the study of these peculiar " monstrosities," that we are enabled to ascertain the actual existence of particular organs in a latent or undeveloped state; and it has been by connecting the results of such inquiries, that the whole theory of the natural classification of plants has of late years undergone a complete revolution. The chief phenomena which regulate the conditions essential to the extension of this kind of knowledge, are the abortion, degeneration, metamorphosis, and adhesion of certain parts. The account of these belongs more especially to the organographical department of botany, and very little is known to the physiologist of the causes which produce them. The non-development, or "abortion" of any latent organ in a plant, seems to arise very frequently from its compression by some contiguous part, or else from an abstraction of its nutriment by another part which exerts a greater vital activity. As these effects depend upon the relative position of such parts, the influencing cause begins to operate even from their nascent state, and long before their form is discernible by us. We have, consequently, no control over these causes, and their influence could never have been noticed by us, if nature herself had not assisted in the discovery by producing those occasional aberrations from the ordinary state of plants which are known by the name of "monstrosities." That all the various parts of the fructification are modifications only of the leaf, is demonstrable by an appeal to numerous examples of monstrosities in which these parts may be seen to possess an intermediate character. But we are still utterly ignorant of the nature of those predisposing causes which are capable of effecting such wonderful modifications in the form, colour, consistency, and nervation of this single organ, and, above all, such a complete dissimilarity between its various functions.

The modifications resulting from the "adhesion" of corresponding parts are of great importance in studying the affinities of different species, and the laws which regulate this phenomenon are of high interest to us in an economical point of view; for it is upon the knowledge of them that a true theory of grafting must depend, an operation no less useful to the horticulturist, than its effects are wonderful to the physiologist. "Adhesion" consists in the perfect union or blending of the cellular tissues of two parts, which are supposed to have been originally distinct in their nascent state; and it is to be met with in every class of plants. Besides the ordinary causes of adhesion, resulting from some constant predisposition in the plants themselves to produce this effect in certain parts of their texture, we often meet with it in nature, where it has evidently been the result of accident. Two fruits, or two branches, by growing close together, will gradually adhere by the union of their tissues, and then continue to grow as one specimen Now grafting is nothing more than the artificial production of the same effect. The art is reputed to be of Phœnician origin; and the Romans have left us many fabulous accounts of the wonders which, as they asserted, might be performed by it; none of which, however, are so wonderful as the simple truth itself. It has been customary, in all works on this subject, to describe the union of the graft and stock as commencing between the libers, or inner layers of bark, in each; but De Candolle exposes the fallacy of this opinion, and has shown that it is their alburnums, or outermost layers of wood, which first unite. It is through the alburnum that the rising sap is conducted to the graft, and the adhesion of the libers cannot be affected until afterwards, when the descending sap returns into the stock by this channel. A graft never succeeds, excepting between such species as are nearly allied, or which at least belong to plants of the same natural family, and no credit is to be given to the accounts of fruit and flowers having become considerably modified by being grafted on plants of an opposite nature from themselves: such as the pretended case of a rose, on a black currant, being rendered black; or of a jessamine, on an orange, obtaining the scent of the latter. The only apparent exception to the law which requires that the graft and stock should be at least of the same family, occurs in certain parasites which are found on trees of different natural orders. The misletoe, which is a seeming exception in many particulars to the ordinary laws of vegetation, is a plant of this kind. But there is this very esssntial difference between the manner in which the misletoe adheres to the tree upon which it lives, and the union of the graft and stock: whilst the latter unite both by their wood and bark, the former unite by their wood alone. This parasite, requiring only a supply of the rising sap, which is much the same in all plants, is afterwards enabled to nourish itself by elaborating this material in its own organs; but then it can restore nothing to the tree upon which it is supported, owing to the want of union between the tissues of their barks. But although, generally speaking, grafting appears to produce no material change in the character of the plant, yet there are certain slight alterations which arise from this cause in particular cases; the size, habit, duration, and even flavour of the fruit being sometimes modified by it. Grafting is, in the economy of vegetation, what commerce is in the political arrangements of society: it creates nothing, but serves as a means of transporting and disseminating that which is most useful and most required. It is of some importance also to the botanist, enabling him to ascertain the affinities of plants in certain doubtful cases.

An ingenious proposal has been made of changing the characters of diœcious plants into monœcious, by grafting a branch from the male upon the female, and thus superseding the necessity of cultivating the former in plantations where it has hitherto been found necessary to do so, merely for the purpose of supplying the fertile individuals with the pollen requisite for setting their fruit.

Another phenomenon to be noticed here, is the constant descent of the root and ascent of the stem, which philosophers had been puzzled to account for. Some had ascribed it to a kind of instinct in plants, analogous to that which is observable in animals; others considered it to arise from a peculiar effect of the vital force. At length Mr. Knight, in the Philosophical Transactions for 1806, demonstrated experimentally, that the proximate cause at least of this phenomenon depended on the law of gravitation. This he proved by causing seed to grow on the circumference of two revolving wheels; one of which was placed vertically, and the other horizontally. The seeds on the vertical wheel germinated with their plumules directed towards its centre, and their radicles from it, which effects could be ascribed only to the centrifugal force having been substituted by this arrangement for the force of gravity. In the horizontal wheel, the radicles were inclined outwards, and the plumules inwards with respect to the vertical axis of the wheel; the action of gravity not being nullified, but merely having its effects modified in this case, by combining with those produced by the action of the centrifugal force. Still, however, a question arises, as to the precise manner in which the force of gravity produces these effects upon the germinating seeds. De Candolle explains this by referring to the difference between the mode of growth in the root and stem. The root increases in length by successive additions to its extremity only, which is continually in a soft state, and always tending to descend vertically by its own weight; but the stem increases for some time through out its whole extent, and if it be at all inclined to the horizon, the grosser and more nutritive particles in the sap tend towards the lower surface, which is in consequence rendered more vigorous than the upper, and the fresh fibres, extending themselves more rapidly on that side, produce an incurvation upwards. Upon a similar principle he explains the tendency of the ascending organs to turn to the light, which, by decomposing the carbonic acid in the young shoots, would necessarily cause that side which is most exposed to its influence to fix most carbon in its tissue; and, consequently, to become sooner rigid than the other side, whose fibres would be more extended, and thus produce the incurvation observed. The causes of the tendril twisting itself into a spiral, and of the convolution of the stems of climbing plants, are unknown. Dr. Wollaston suggested that these effects might depend upon the diurnal course of the sun, and proposed that plants should be raised in the southern hemisphere from the seeds of certain climbers which had grown in the northern hemisphere, in order to see whether the direction in which the convolutions would take place, might not be different from what it was before; but it does not seem to us evident why any such result should be at all expected from this experiment, and we have climbers from the southern hemisphere turning the same way as other species of the same genus which grow in the northern.

One great distinction between animals and vegetables consists in the power which the former possess of spontaneously moving various parts of their bodies; for, although there are certain plants which are known to have a power of motion in some of their organs, yet this is very different in its character from that which animals exhibit. The sleep of plants, as it is termed, is a phenomenon of this class. Not only do many flowers close and expand at different hours during the day and night, but the leaves also of many plants exhibit a similar property. This is extremely common in the compound leaves of the *Leguminosæ*, whose leaflets close together in pairs, either by their upper or under surfaces. Eleven distinct modes have been observed in which leaves fold together in the sleep of plants; but in all these cases, the sort of effect which takes place is very different from that which is manifested by the sleep of animals. It is a mere change of position, without any relaxation of the parts affected, which remain equally rigid as before. The action of light evidently produces an effect in determining and modifying the sleep of plants, though it does not appear that this agent can be regarded otherwise than as a stimulant, and not as the efficient cause of the phenomenon, which seems to reside in a predisposition of the plant itself to assume these periodical changes. A singular example is noticed of a species of *Acacia*, whose leaves sleep by night, and its flowers by day. In a few species, similar motions may be excited in these organs, by the application of mechanical or chemical stimulants, of which the Sensitive-plant affords the best known and most remarkable instance.

The different colours in plants are probably owing to different degrees of oxygenation in their chromule, which is every where present in their cellular texture in the forms of minute globules. It is colourless in its primitive state; but as soon as the parts capable of becoming green are enabled to decompose carbonic acid by their exposure to light, the chromule assumes a green tint. In the autumn, when oxygen continues to be absorbed, but ceases to be given out again by the leaves, the chromule assumes a yellow or a red tint. As the small globules contained in the cellular tissue of the flower and other coloured parts appear to be of the same nature as those in the leaf, it is the most plausible conclusion to suppose, that all vegetable colours are modifications of this single substance. A very small portion of light seems to be all that is necessary for colouring the chromule of some plants. Humboldt observed à sea-weed (*Fucus vitifolius*), brought up by the lead from the depth of one hundred and eighty feet, to be perfectly green, though the light diffused at the spot where it had grown must have been nearly fifteen hundred times less than at the surface of the sea. It is almost a general rule, that those species which possess varieties of a blue colour, do not embrace varieties of a yellow colour, and *vice versâ*, though either class may possess others which are red or green. The following scale will represent what other varieties may, *à priori*, be expected in a given species, upon observing a variety which possesses one or other of the colours mentioned in the right or left-hand column :—

GREEN.

Greenish blue.	Greenish yellow.
Blue.	Yellow.
Violet-blue.	Orange-yellow.
Violet.	Orange.
Reddish violet.	Orange-red.

RED.

There are, however, several decided exceptions to this rule; and it does not hold good in some party-coloured flowers, as in the *Convolvulus tricolor,* which has a yellow zone in the throat, and a blue band at the summit of the corolla.

All white flowers are exceedingly diluted tints of different colours; a fact which first furnished a valuable hint to the flower painter Redouté, who arrived at very great perfection in the art of representing white flowers on a white ground. All blacks are intensely deep tints of various colours.

Nothing is known of the causes which produce so great a variety in the taste and scent of different plants. The senses are bad criterions for enabling us to decide and classify these phenomena, as we may readily perceive when we find the odour of arsenic to be exactly like that of garlic, and the smell of musk to exist in several plants. It is observed that the vegetable substances which are the most tasteless are frequently those which afford the most nutriment, and hence it is customary to improve their flavour by the addition of some condiment. This result is often obtained naturally in certain alimentary species, which are nearly allied to others of a dangerous quality, or is produced by blanching some poisonous plants themselves, when their bad qualities are masked by the superabundant quantity of wholesome nutriment that is thus formed.

The discussion on the " individuality" and "duration" of plants is of great interest; involving, as it does, questions of national importance respecting the improvement and growth of timber plantations, and connecting the inquiries of the botanist with those of the geologist respecting the previous conditions of the earth's surface.

Various opinions have been broached as to what ought to be considered an " individual" in botany; some would apply the term in its ordinary acceptation to every separate plant, whilst others consider each separate bud as a distinct " individual;" and others, again, would have every cell of the cellular texture to be elevated to the rank of an " individual" being. De Candolle allows each of these suppositions to maintain its ground, by calling a single cell " an individual cell," a single bud " an individual bud," &c.; whilst he himself proposes to consider the duration of an " individual plant," without caring whether it may have originated from the development of a cell, seed, tubercle, slip, or from any other source. There are two modes of considering an " individual plant :" either as an assemblage of buds round a common axis, where each bud is supposed to be endowed with a separate existence; or else as a single being, whose functions are performed by certain organs, the whole of which are annually replaced by a completely new set. Either hypothesis will allow for the life of an individual plant being indefinite in duration. If we consider the trunk of a tree to be increased by the materials accumulated by fresh crops of buds annually produced upon its surface, then it bears a strong analogy to a coral reef, the animals of which possess an individual existence, and are separately employed in increasing the aggregate mass of their habitations. Strictly speaking, then, the death of a tree can never result from any effects of old age at all similar to those which necessarily destroy life in the animal kingdom, where we find a very limited period of existence assigned to each individual, in consequence of the obstruction of those organs which are destined to perform the function of nutrition, which may be carried on during the old age of the tree as vigorously as it was performed by the young plant. Those causes, then, which ultimately destroy life in plants must be classed as accidents, or as proceeding from various diseases, induced by the influence of external agency; large limbs are broken off by their own weight, and thus rottenness is introduced into the heart of the trunk, which gradually becomes too feeble to support the foliage, and is blown down. But wherever these and similar accidents have been prevented, trees have attained to a vast antiquity, and there are very plausible reasons for believing that there are at this time in existence many which have endured far beyond the records of history, and must have been standing shortly after the last general catastrophe to which this earth has been subjected. That the life of very many plants is necessarily short, as in the case of annuals, biennials, and others, seems to arise from the complete exhaustion which they suffer during the maturation of their seed, all the nutriment prepared in their stems being wholly abstracted by this effort.

In order, however, to give some weight to these conjectures respecting the possible duration of certain trees, it is necessary that we should point out the method by which we are enabled to approximate to the age of very old trees, with some tolerable degree of certainty. In many exogenous trees, which is the character of all the timber of temperate climates, the number of concentric zones observable in a transverse section of their stems, affords an exact measure of their duration, provided the section be made near the root. By placing a strip of paper upon the surface of such a section, and extending it from the centre to the bark, the distances between the several zones can be marked upon it, and thus a register may be formed both of their number and of the relative growth of different years. On account, however, of the frequent inequalities in different parts of the same zone, it is better to take the girth of the tree, and obtain the mean rate of increase, by dividing the mean diameter by the number of zones. No good result can be obtained from any observations of this kind upon trees that are much below a hundred years, as their rate of growth is too unsteady, and varies too much in different individuals prior to this period; but very useful averages may be obtained from old trees, because, after a certain age, they obtain a more settled rate of increase. The averages which are thus obtained, will serve us for approximating to the age of others. They may also serve as a test for calculating the relative worth of timber of the same kind, as a building material : since the preserving quality in wood depends upon its compactness, and this again upon the slowness of its growth, it may be seen, by simply inspecting the layers of any particular specimen, whether its age is above or below the general average of trees of the same bulk, and consequently whether the compactness of the timber is greater or less than usual. There are various methods of obtaining a scale, which may serve for approximating to the ages of trees, besides the one just mentioned. Their rates of increase may be obtained by measuring their trunks at successive intervals of time; or a lateral incision may be made, and the number of layers counted to a certain depth. In all these expedients, however, the observer must be careful to make great allowance for the fact, that trees increase more rapidly in the early stages of their growth than afterwards. The dimensions of several very large trees have been recorded both by ancient and modern observers, and various conjectures have been offered respecting their probable ages. Some of these trees, indeed, like the celebrated chesnut of Mount Ætna, appear to have resulted from the union of several trunks which had grown near together. There are others, however, as the oaks, and more especially the yews, recorded by Evelyn, which are single trees of vast antiquity. De Candolle, by computing the results of several observations, has ascertained the average increase of the yew to be about one line, or the twelfth of an inch, in diameter, yearly. Applying this rate for calculating the ages of the four most celebrated yew-trees in Great Britain, whose dimensions are on record, he finds them respectively to have lived 1214, 1287, 2558, and 2880 years. In the first of these examples, we have the testimony of history for knowing that this tree was in existence, and must have been of considerable size, in the year 1133, it being recorded that the monks took shelter under it whilst they were rebuilding Fountain's Abbey. These and other facts respecting the probable duration of some of the largest European trees, throw considerable plausibility on the views of Adanson, who, nearly a century ago, had constructed a table, from a regular geometrical formula, for calculating the probable ages of the enormous Baobabs of Senegal. The extended duration of these trees is favoured by the circumstances of their not attaining to any great height, and by their growing in a country where they are never exposed to the effects of frost; there are several examples of their trunks attaining to the enormous dimensions of sixty, and even ninety feet in circumference. Adanson mentions the data upon which he proceeded in constructing his table and there is no apparent reason for our supposing that his conclusions do not lie within the truth. For example, his table ascribes the age of 210 years to a trunk six feet in diameter: but he had found some trees of this size in a small island off Cape Verd, upon which he noticed the traces of inscriptions, some of which were dated from the fourteenth, and others from the fifteenth century.

Now we can hardly suppose their diameters to have been less than four feet at the time when the inscriptions were first carved upon them : a supposition which allows an increase of only two feet in 300 years, and which would consequently make them 800 instead of 210 years old, as shown by the table. What, then, must be the age of a similar tree of thirty feet in diameter? The table itself ascribes to it a life of 5150 years! But this subject has hitherto engaged so little of the attention of observers, that we want additional testimony before we can be expected to place much confidence in speculations which, it must be confessed, are, at first sight, very startling. De Candolle, indeed, seems to consider the question in some measure settled, and sums up his account with the following remarks:

"I think that I have given a detailed proof in this section of the existence, past or present, on the earth, of some very old trees, viz. of an

Elm	. . . . of	335 years
Cypress	. . . . . about	350
Cheirostemon	. . . . about	400
Ivy	. . . .	450
Larch	. . . .	576
Orange	. . . .	630
Olive	. . . .	700
Oriental Plane	. . . . . .	720 and upwards
Cedar of Lebanon	. . . about	800
Oak	. . . .	810; 1080; 1500
Lime	. . . .	1076; 1147
Yew	. . . .	1214; 1458; 2588; 2880
Taxodium	. . . . . . about	4000 to 6000
Baobab	. . . .	5150 (in the year 1757)"

—p. 1007.

Before we quite dismiss these wonders, we must mention that M. de Candolle appears to have somewhat exaggerated, or, as some may think, improved upon, the account of the Baobab given by Adanson in his "Familles des Plantes." That excellent observer stated the inscriptions which he examined to have been on the *surface* of the tree, but M. de Candolle has somehow made out that he had detected them in the *inside!*

"The Baobab," says he, "is the most celebrated instance of extreme longevity which has hitherto been noticed with any degree of accuracy. In its native country it bears a name which signifies 'a thousand years;' and, contrary to what is usual, this name expresses what is in reality short of the truth. Adanson has noticed one in the *Cape de Verd Islands* which had been observed by *two English travellers* three centuries earlier: he found *within* its trunk the inscription which *they* had engraved there covered over by *three hundred woody layers*, and thus was enabled to estimate the bulk by which this enormous plant had increased in three centuries."!!—p. 1003.

Let us compare this account with Adanson's own words, observing the passages we have noted in italics.

"Those which I saw in 1749 on the ' *Isles de la Madeleine*,' near

Cape Verd, with inscriptions of *Dutch* names, such as Rew, and other, *French* names, the former dating from the fourteenth and the latter from the fifteenth century, which inscriptions I renewed, merely adding below them, ' renewed in 1749,' were then about six feet in diameter. These same trees were seen in 1555, that is to say, two hundred years earlier, by Tevet, who notices them in the account of his *Voiage aux Teres Antarktikes*,* describing them merely as ' fine trees,' without mentioning their thickness, which must at least have been three or four feet, judging from the little space occupied by the characters forming the inscription ; they had therefore enlarged about two or three feet in *two hundred* years."—*Familles des Plantes*, Preface, p. ccxvi.

If, again, we compare these passages with De Candolle's former observation on the subject, in his *Organographie*, we confess ourselves at a loss how to account for his present mistake, otherwise than by supposing it to have arisen from his trusting merely to memory, which in this, as well as in some other instances we have marked, has been treacherous.

" Whenever we find on old barks the traces of some ancient inscription, it may be made use of as an index where to search for it in the corresponding part of the wood, and then, if it has originally penetrated to the alburnum, traces of it will be found buried beneath the woody layers : in this case an exact verification may be obtained of the age of the inscription, and of that of the tree. *If* Adanson could have done this on the Baobab of the ' Madeleine,' we should be in possession of a more certain document of the real age of these veterans of the organized world."—*Organog.* vol. ii. p. 192.

There is no direct method by which we can ascertain the age of endogenous trees, and as these are chiefly confined to tropical climates, botanists are as yet in possession of very few documents on this subject. The celebrated Dragon's-blood (*Dracæna draco*) of Orotava, in Teneriffe, appears to be an individual of prodigious antiquity. When the island was first discovered, in 1402, this tree was held in veneration by the natives for its great size ; and the four centuries which have since elapsed had seemingly produced very little alteration in its general appearance up to the year 1819, when a large portion of the top was blown down in a violent tempest : the remainder, however, still continues to flourish in its pristine vigour.

III. We have already seen that, in order to comprehend the phenomena connected with the two grand functions of Nutrition

---

* The title of Tevet's work is *Les Singularitez de la France Antarctique*, 1 vol. 12mo. Anvers, 1558, and the account alluded to occurs at p. 18, accompanied by a rude woodcut of this *arbre estrange*.

and Reproduction, it is necessary to have recourse to experiment, and to examine the various results produced on vegetation by the influence of certain external agents, as light, heat, &c., under different circumstances. When we have examined into the several causes which may have tended in any way to modify the action of the vital principle, and have thus learnt to appreciate the precise share which each has borne in producing the whole effect, we may then separate these results into two classes, the one embracing such as may be ascribed to the action of the vital principle alone, and the other including such as are due to the operation of the external agents. The details already attempted properly belong to the first class; whilst those of the second form a distinct subject, under the newly invented title of " Vegetable Epireology."* The presence of some of these external agents, as light and heat, is essential to the condition of life itself; and therefore the only effects produced by their agency, which require further examination, are those in which their influence is exerted in a greater or less degree than is essential to the healthy state of the plant. Others, however, no way essential to its life, are nevertheless instrumental in modifying its form, or in injuring its health. Surrounded as all organized beings are by an assemblage of various matters, some useful and some noxious to them, their life consists in a constant series of attempts to profit by the presence of the one kind, and to repel the attacks of the other; and as plants are less provided with a variety of resource for accommodating themselves to existing circumstances than animals, we might, *à priori*, expect to find them subject to greater modifications. It is this part of our subject which furnishes the data upon which our speculations in the geographical and agricultural departments of the science must depend, and upon which also we must frame our notions of Vegetable nosology. Epireology may be considered as the counterproof of physiology, the latter teaching us the effects produced by the living plant on the external agent, and the former the effects produced by the agent on the plant. The *first* of these agents is " light," whose action is that of an excitant. If it be too strong for the nature of the plant exposed to its influence, the several functions of nutrition, viz. the succion of the spongioles, the exhalation of the leaves, and their decomposition of carbonic acid, are performed in excess, and a proportionate defect accrues in perfecting the seeds, few or none of which are then ripened. If the light be too feeble, these necessary functions are not completed, and the plant

---

* Επιρροη, influxus; vel επιρρεω, increpo.

becomes pale and dropsical. Many effects which are popularly ascribed to the influence of the atmosphere, are in fact due to the action of light. Thus the more robust condition of such trees as grow on the outskirts of a forest must be attributed to the greater light which they enjoy; and thus there is an increased produce from corn which is grown upon ridges alternating with others that are allowed to lie fallow. Stove plants in our northern climate suffer much from a want of light in winter, and an ingenious suggestion has been thrown out for remedying this defect by supplying them with gas-light.

A *second* external agent is " electricity," whose action, however, upon vegetables is so little understood that it need scarcely be alluded to. It seems to operate as a stimulant; and it has been observed that vegetation is more vigorous in stormy weather, and that rain is always more serviceable than any artificial irrigation, circumstances which possibly may be ascribed to the electrical conditions of the atmosphere at these times.

A *third* agent is " heat," which also acts as a stimulant to vegetation: an increased temperature augmenting the suction of the roots, determining and accelerating the germination of the seed, the period of flowering, the maturation of the fruit, &c. &c. Different degrees of temperature are suited to the constitution of different species, but the effects which an increase of temperature produces on the " excitability" of plants are not so well understood as those which it produces on the various materials of which the plant is composed, and the several substances provided for its nutrition. These effects are less decided upon the solid parts than upon the liquids contained within them, and hence the plant is rendered more capable of resisting the extremes of heat and cold in proportion as it contains less water or possesses a more viscid kind of sap. But independently of all the physical causes by which we would explain the influence of temperature on vegetation, there is certainly a specific " excitability" peculiar to different species, by which they are better adapted to live in one temperature than another, and by which, therefore, the natural limits of their several ranges on the surface of the earth are fixed. The effect of an excess of temperature upon the health of a plant varies according as it is accompanied by drought or by humidity. In the first case the plant has a greater tendency to produce flowers, and in the latter leaves. Where the temperature is too low, vegetation becomes languid—a fact which affords us a useful hint not to sow seeds too early, since the stronger plants, which are produced by the warmer season, make up for the lateness of their germination by the greater vigour of their growth. The important subject of the " naturalization" of plants depends chiefly on considerations connected with the effects produced by different temperatures upon their several constitutions. It is not merely the mean annual temperature which is to guide us in these inquiries, but we must also be watchful of the circumstances connected with the inequalities in its distribution; for it is the extremes of heat and cold at a given place which are found to be most influential in fixing limits to the range of each species in its natural condition. We must also remark, that the laws of agricultural geography will differ from those assigned to botanical geography, inasmuch as there are many species which may be cultivated with success in places where the function of nutrition may be carried on, but where they have not the power of completing the function of reproduction; and then man may continually import fresh seed, and so obtain a constant supply of the young plant. Many erroneous opinions have been advanced on the subject of naturalizing plants, and it has been imagined that, by obtaining a succession in the culture of any species at the same spot, it might gradually be made to sustain a much greater or much less degree of heat than it could support in its native country. It is quite clear, however, that the real fact has been very much overrated, and that the limits within which a plant admits of being " acclimatized " are generally very confined.

A *fourth* agent is the " atmosphere," whose influence has frequently been confounded with that of light. The homogeneity of its chemical composition in all parts of the world excludes the idea of its having any effect in producing those differences of climate which exist under the same parallels of latitude; but these effects must principally be ascribed to the several degrees of moisture with which it is charged in different places. Thus, where fogs prevail, the chances of sterility are increased from the moisture injuring the pollen. It is only by means of the various ingredients accidentally mingled in the atmosphere that this agent produces any striking alteration on the general conditions of vegetation. The deleterious influence of the sea breeze is well known; a smoky atmosphere is also injurious, but the quantity in which various gases are to be met with in it is generally too minute to be productive of any appreciable effects, and we must look to its mechanical and physical properties for producing the greatest influence in modifying the ordinary conditions of vegetation. The agitation of the atmosphere, when not too violent, is decidedly serviceable; its diminished density in high altitudes does not appear to produce any material influence; but this is very difficult to be calculated, as it is connected with other effects, which result from the increase of light and the diminution of temperature and humidity.

A *fifth* external agent is " water," whose presence is so very essential to vegetation. The effects which result from a deficiency in its supply are flaccidity and death: an excess produces an increased tendency in the buds to run to leaf instead of to flower, and the whole plant becomes inclined to rot. Although the basis of all agricultural speculations for a given district ought in some measure to be formed with reference to the mean quantity of rain which annually falls there, yet it is equally necessary to take into consideration the periods of the year in which the rains are most prevalent. Thus the autumnal rains of the south of France are more influential in excluding the naturalization of the cotton-tree from those parts, than any deficiency in the temperature of the climate.

A *sixth* agent is the " soil." However strongly it may have been asserted that the soil has no direct influence on vegetation, because it can be shown that plants may be raised without its assistance, yet it is manifestly of too great importance, both physically and chemically, in the ordinary process of nature, to be passed by unnoticed. It seems, indeed, to be clearly established that this influence is chiefly to be ascribed to the variable degrees in which different soils retain their moisture; but it is not so clear that the earthy particles composing them, when introduced into the system with the water absorbed by the spongioles, produce no effect whatever upon the plant. The alkalis most certainly produce a very marked influence, though even here there are some plants which seem to be indifferent as to which of them they may happen to be supplied with. We think M. de Candolle advances a singularly untenable hypothesis when he suggests that maritime plants do not require the presence of salt as essential to their healthy condition, but may rather be considered in the light of individuals which are better capable of repelling its deleterious effects than others of a less robust constitution. It is true that maritime plants will not live where salt is in excess, any more than others; but this is no argument for supposing that the precise supply which they receive in their natural state is not essential to their perfect health; otherwise, why are so many of them limited to the sea shore, and why do some of them re-appear inland in the neighbourhood of brine-springs and salt-pits? In general we may allow that the direct influence produced on vegetation by the chemical composition of the soil is extremely trifling, and that the greatest effects are due to the mechanical and other physical properties which it possesses, such as its mobility, tenacity, colour, &c. &c. It is over these peculiarities that man possesses the greatest control, and where he is often enabled by his art to bring under culture extensive districts which otherwise would continue for ever bare of vegetation. No districts of country appear so hopeless of being reclaimed from utter barrenness as those which are subject to a periodical invasion of sand, blown over them by the prevalence of certain winds. The sands on the sea shore, from whence this supply in general originates, by their capillary action attract and preserve in their interstices the water which may chance to be beneath them, or perhaps the soft portion of the sea water itself, and consequently they are always somewhat moist at a little distance below the surface, though the extreme dryness of the surface itself prevents any vegetation from fixing itself upon it. In Holland, England, and other countries where this is the case, it is usual to plant, and to preserve with great care, the sea grasses and sedges (*Arundo arenaria*, *Elymus arenarius*, *Carex arenaria*, &c.), which possess long creeping rhizomata, sufficient in some measure to bind the sand, and check its being drifted by the wind. But this expedient is very limited in its effects, and not to be compared with the benefits that have resulted from planting timber in the manner that has been practised on the coast of Gascony, where it was first adopted by an engineer of the name of Bremontier.

"Bremontier's plan," says De Candolle, "is wonderful for its great simplicity. He sows, in the dryest and most shifting sand, the seeds of the broom (*Genista scoparia*) mixed with those of the sea-pine (*Pinus maritima*), and then covers over the spaces that are sown with branches from the nearest pine forests, by which means the sand is, to a certain extent, prevented from shifting. The broom springs up first, and thus serves the double purpose of further restraining the sand, and of nursing the young pines. The latter grow for seven or eight years under shelter of the broom, whose foliage becomes mingled annually with the sand, which it thus partially fertilizes. After this period the pine overtops the broom, and frequently entirely kills it with its shade. In ten or twelve years the rising forest is thinned for the manufacture of tar, and for procuring branches to cover the newly sown districts. After twenty years have passed, a fall of the trees commences for the manufacture of resin. These forests, placed on the dunes (drifting sand-hills) along the sea side, shelter the whole country behind them from the continuous action of the westerly winds, and thus, whilst they themselves yield a supply of an important article of commerce, they protect the produce of the rest of the country. It is highly desirable that this prodigious undertaking, the most splendid agricultural enterprise of our age, should gradually be completed, and thus provide a shelter for the whole district between the mouths of the Adour and Garonne. I regret that the character of this work does not admit of my entering into further detail, and I close this account by stating that I have herborized during a whole day in these forests sown by Bremontier on perfectly dry sand, upon which, before his time, there could scarcely be seen any trace of vegetation."—p. 1236.

It is difficult to estimate the influence which the chemical action of an earthy substance may be supposed to exert on plants, since all soils contain nearly the same ingredients in various degrees of intermixture. Any botanic garden is sufficient evidence how large a number of different species may exist in the same soil. The various methods adopted for rendering soils more fertile must depend upon the different modes in which each particular soil is observed to affect the plants that grow upon it. De Candolle devotes a chapter to these inquiries, and estimates the effects produced by manual labour, the addition of stones, of sand, of clay, &c. &c., and describes the process and result of manuring with decomposing organized matters. As these details are somewhat foreign to the specific object of his work, we shall not dwell upon them, but pass on to a subject more strictly physiological in some of its details, and of nearly equal importance in the practical results which it involves—we mean the process employed by nature in healing whatever wounds trees may receive externally. It is upon this that the whole success of pruning rests, and we shall the more willingly dwell upon the subject, as it is one which has lately given rise to considerable discussion, it having become a matter of doubt among some extensive landed proprietors of this country, whether the old system of " close pruning" has not been productive of immense damage to our forests, and whether it ought not to be entirely laid aside in favour of another system, which has been termed " fore-shortening." These inquiries appear to have originated from an examination of the effects which are always produced in timber whenever a branch is cut off close to the trunk of the tree; and it is rightly asserted, that wherever this had been done, a complete solution of continuity exists between the old wood and that which had been formed over the wounded surface; and that in many cases, though by no means in all, decay and rottenness has been introduced to a greater or less extent into the very heart of the tree. Where the pruned branch is large, the blemish thus introduced is proportionably great, and must always diminish the value of the timber, even though it should not have caused the introduction of any rottenness. But although such, it must be allowed, are undeniably the bad effects of close pruning, it does not follow that the proposed alternative of fore-shortening is any way preferable to it; and we shall first consider the mode in which all wounds are healed over, and then examine the question whether the damage produced by fore-shortening is more likely to be counteracted by the subsequent efforts of vegetation than those which are introduced by close pruning.

Whenever the stem is wounded by the removal of a portion of the bark, and the wood becomes exposed to the action of the atmosphere, a double influence is exerted in gradually decomposing its texture: one arises from the union of its carbon with oxygen, as in the ordinary process of decomposition; and the other from the humidity by which it is penetrated, dissolving some parts of the tissue, and in reducing it generally to a soft and disorganized state. These results will of course be very different according to the nature of the wood attacked, and to the length of time that the wound requires to be healed over. As soon, however, as the wood is secured from further exposure to the atmosphere, the damage ceases to increase, though the blemish which has been introduced admits of no remedy. The new wood and bark which form over the wound, are derived from the growth of the alburnum and liber, which gradually extend themselves from its upper edge, and from along each side, till they meet in the middle and then unite and blend together as in the case of grafting. By judiciously splitting a block vertically at the zone which corresponds to that year's growth in which the surface of a pruned branch was covered over by the fresh wood, every mark of the pruning knife will be found on the discoloured surface of the old wound, as fresh as when it was first impressed upon it, and the new wood will have received a reverse impression of this surface as accurately as a counter receives the stamp of a die. Vertical wounds on the surface of the trunk are those which heal the most readily, because their direction tallies with the course of the cambium, which soon forms a tumour at the upper extremity of the wound and down each side, in the manner just described; and this is more readily extended over the wound in proportion as its surface is smoother. Various composts are useful for protecting the exposed surface from the atmosphere, whilst the healing process is in progress: but nothing of an oily or poisonous description should be employed. Whenever, therefore, pruning is absolutely necessary, it is advisable to prune close, in order to reduce the exposed surface to the condition of a vertical wound; unless, indeed, the limb be very large, when it may be more advisable to prune at some little distance from the trunk, lest the blemish which would be introduced into the timber should be so considerable as more than to counterbalance any advantage that would be obtained. There is no direct means by which a transverse section through the wood may be healed over, and if a branch be lopped at a distance from its point of union with a main branch, or with the trunk itself, the exposed surface never heals over, but causes the decay and death of the branch for some distance back, until this is stopped at some spot where the returning sap is in sufficient quantity to produce fresh wood and bark. The system of " fore-shortening" rests upon the gradual decay of the pruned branch, until it be ultimately killed by the increasing shade of the superior branches, when its fall will take place in the natural way; as in all branches which grow low upon the stem, and are early stifled by the shade of the upper branches, and which slough off, without producing any very marked blemish in the heart of the tree. Here, however, we must observe, that there is no process for "sloughing off" the decayed parts of vegetables which at all resembles that which takes place in animals; but when the branch has become so completely rotten, as to fall off upon the application of the slightest force, it will be found that the new bark and alburnum which are formed round the base of its stump, always envelops more or less of the rotten wood, which forms a rough and jagged surface to the wound. It is erroneous to suppose that those branches which fall off by a sort of natural pruning, resulting from their being killed by an obstruction of the light, leave comparatively little or no trace of their decay in the heart of the tree; but since it happens that those branches which perish early are always proportionably small, when compared with the bulk to which the trunk attains, the blemishes which they leave may easily be underrated, and this we believe to have been the origin of the error which supposes that the blemish introduced upon the natural decay and fall of a branch is, *ceteris paribus*, of less consequence than that which results upon closely pruning it. The danger which attends all pruning may be diminished by paying attention to a few rules, such as cutting the surface quite smooth, cutting it obliquely so as to prevent the wet from lodging upon it, and especially by cutting close to the main branch or stem. The main object is to procure a rapid development of the new wood, in order that the exposed surface may be secured as speedily as possible from the action of air and moisture; and this, we believe, is best obtained by reducing the cut as nearly as possible to the condition of a vertical wound on the stem. All pruning, then, should be avoided as much as possible: but where it is absolutely necessary, it should be performed as soon and as completely as the young plant or branch may bear it with safety.

We have yet to detail the effects produced by poisons on the vegetable structure, and to refer to some of the important practical results which follow from their consideration. The action of poisons on vegetables is analogous to that which they produce on animals. One class is corrosive, and destroys the tissue on which it acts; whilst another class is narcotic, and destroys vitality without producing any decided alteration on the tissue itself. It has been ascertained that nearly all substances which are poisonous to animals, are likewise so to vegetables, though the intensities of their several actions are different in the two kingdoms; but, besides these, there are many substances innoxious to animals which are destructive to vegetable life. In fact, it should seem that almost every thing that vegetables can imbibe is injurious to them, excepting water, the insipid earthy salts, carbonic acid, and other gases, gums, and mucilaginous substances, and finally, certain animal matters when introduced in very weak solution. It has been supposed that the presence of a nervous system might be assumed to exist in vegetables, from the mode in which they are destroyed by narcotic poisons; but there is this remarkable difference in the mode in which these substances act on animals and on vegetables: on the former they act by "sympathy" upon certain parts with which they have no immediate contact, whilst in the latter they produce their effect only on those parts of the tissue into which they are introduced. In vegetables, also, all poisons exert their action upon the cellular tissue, whilst in the more complicated structure of the animal frame different poisons will attack only particular tissues; which again seems to prove the existence of no more than one single faculty in vegetable life, as we concluded to be the case, from other considerations, in the beginning of this article. It is a curious fact in the action of vegetable poisons, that a plant may be killed by the poison which it has itself secreted, as a viper may be stung to death by its own venom. Hence it has been very generally noticed, that the soil in which any particular plant has grown, and into which it has consequently discharged the excretions of its roots, is rendered noxious to the growth of plants of the same or of allied species, though it be quite adapted to the support of other species. This fact is of the greatest importance in an economical point of view, as the whole theory of the rotation of crops may be considered to depend upon it. The discovery of this important step in agriculture was probably made by the Belgians; at least they have the merit of having developed the theory of it. Formerly it used to be said, that the whole secret of good husbandry consisted in ploughing well, and in manuring well; but to these must now be added the equally important art of so arranging the cultivation of different crops that they may mutually assist each other, and thus enable the farmer to obtain the greatest possible annual return from the same land. The whole theory depends upon the fact, that all plants succeed badly upon lands which have lately borne crops of the same species with themselves, or even of the same genus, or of the same family. This effect is not owing to any exhaustion of the soil that must have taken place during the growth of the previous crop, but arises from a corruption of the soil, by the intermixture of vegetable excretions given out at the root, which excretions are always more deleterious to plants of the same kind than to others. It is even ascertained that the excretions of some plants are beneficial to the growth of others of a different family; the *Leguminosæ*, for example, improving the soil for the *Gramineæ*. Agriculturists have proposed various theories to account for the beneficial results obtained by a rotation of crops. Some have supposed that one species, by its denser foliage, chokes the weeds which otherwise would spring up, and assists the crop in exhausting the soil; others have attributed the improvement that has taken place to the remains of the previous crop, which they suppose may have served as manure; a third have said, that the roots of different crops extend themselves to different depths, and so extract their nourishment from portions of the soil which do not interfere with each other; and lastly, it has been urged, that plants of different families may possibly derive their nourishment from different materials. It may be true that some of these causes have a certain degree of influence in determining which may be the most proper plants for a rotation, but they can only be considered as of very secondary importance when compared with that which relates to the deterioration of the soil, by its intermixture with the radical excretions of a previous crop. After enumerating some of the collateral circumstances which should direct the judgment of cultivators in selecting such plants as may be best adapted to a rotation of crops in any particular district, De Candolle proposes the following fundamental and physiological principles, which ought to be attended to where complete success is to be expected. *First*, a new crop ought never to succeed another of the same kind, unless under some very peculiar circumstances, as where the soil is annually renewed, or where it is naturally so fertile as to be capable of resisting the inconveniences which ordinarily result from such a system. *Secondly*, a new crop ought not to succeed another which has been raised from plants of the same family. A remarkable exception to this rule occurs in the practice adopted in the valley of the Garonne, where the soil admits of a biennial alternation between wheat and maize. *Thirdly*, all plants with acrid and milky juices injure the quality of the soil, and their remains should never be buried after the removal of a crop. *Fourthly*, plants with sweet and mucilaginous juices improve the soil for others of a different family. The chief of these are the *Leguminosæ*, which are commonly adopted in practice for this purpose.

The great importance of this subject may well excuse our author for having entered somewhat more into its details than a work devoted to vegetable physiology might otherwise have warranted.   But botany and agriculture are like two provinces of the same empire, which are separated by a broad river, with theory on the one side, and practice on the other; numerous bridges ought therefore to be constructed across this river, and our author has succeeded in erecting some, and in rebuilding others, on better principles than those which have hitherto been adopted.   It now becomes the duty of the agriculturist to take advantage of them, and to study botany more zealously than he has hitherto done, and perhaps than it was possible for him to do, whilst the descriptive department of the science was still restricted within the limits of an artificial system, and its physiology was entirely based upon vague hypotheses.

An Appendix is added to the work, for the purpose of pointing out to those who may be desirous of rendering their assistance towards the further elucidation of the subject, how they may best accomplish this object.   There are many points of first-rate importance in the establishment of a correct theory, which are as yet undetermined ; so that any one who chooses to enter on this field may soon expect to find ample opportunity for making fresh discoveries.   Not only the descriptive botanist, but the chemist, the natural philosopher, the agriculturist, the distant traveller, and the physiologist, are all called upon to lend their aid in determining certain questions within the sphere of their respective observation, and we cannot possibly do better than close this long article by seconding the wishes of our author, that they may be persuaded to listen to his advice.

TRANSACTIONS

OF THE

CAMBRIDGE

PHILOSOPHICAL SOCIETY.

ESTABLISHED November 15, 1819.

VOLUME THE FIFTH.

CAMBRIDGE:

PRINTED BY JOHN SMITH, PRINTER TO THE UNIVERSITY:

AND SOLD BY

JOHN WILLIAM PARKER, WEST STRAND, LONDON;

J. & J. J. DEIGHTON: AND T. STEVENSON, CAMBRIDGE.

M.DCCC.XXXV.

IV. *On a Monstrosity of the Common Mignionette.* By Rev. J. S. Henslow, M.A. *Professor of Botany in the University of Cambridge, and Secretary to the Cambridge Philosophical Society.*

[Read *May* 21, 1832.]

Having met with a very interesting monstrosity of the common Mignionette (*Reseda odorata,*) in the course of last summer (1831), I made several drawings of the peculiarities which it exhibited. I beg to present the Society with a selection from these, as I think they may both serve to throw considerable light upon the true structure of the flowers of this genus, which is at present a matter of dispute among our most eminent Botanists, and also tend to illustrate the manner in which the reproductive organs of plants generally, may be considered as resulting from a modification of the leaf.

It is well known to every Botanist, that Professor Lindley has proposed a new and highly ingenious theory, in which he considers the flowers of a Reseda to be compounded of an aggregate of florets, very analogous to the inflorescence of a Euphorbia. Mr Brown, on the other hand, maintains the ordinary opinion of each flower being simple, and possessed of calyx, corolla, stamens, and pistil. I shall not here enter upon any examination of the arguments by which these gentlemen have supported their respective views, but will refer those who are desirous of seeing them to the " Introduction to the Natural System of Botany, by Prof. Lindley," and to the " Appendix to Major Denham's Narrative, by Mr Brown." My present object will be little more than to describe the several appearances figured in plates 1 and 2.

Fig. 1. is one of the slightest deviations that was noticed from the ordinary state of the flower. It consists in an elongation of the pistil (*a*), and a general enlargement of its parts, indicating a tendency in them to pass into leaves. This is accompanied by a slight diminution in the size of the central disk. The number of the sepals was either six or seven.

Fig. 2. is a portion of the ovarium of the same flower opened, in which three of the ovules are somewhat distorted.

Fig. 3. Here the three valves of the ovarium have assumed a distinctly foliaceous character (*a*); the same has happened to some of the stamens (*b*), and to the petals (*c*); but the sepals are unaltered. The central disk has entirely disappeared.

Fig. 4. This is a still closer approximation to the ordinary state of a proliferous flower bud, when developed. Those parts which would have formed the pistil, if the flower had been completed, are no longer distinguishable, and only a few of the stamens are to be seen, disguised in the form of foliaceous filaments crowned by distorted anthers (*b*).

Fig. 5. A slight deviation in one of the petals from the usual character. The fleshy unguis is somewhat diminished, and the fimbriæ are becoming green and leaf-like. These are aggregated into three distinct bundles, the middle one being composed of a single strap, and the two outer ones of five straps each, blended together at the base.

Fig. 6. The line of demarcation between the unguis and the fimbriæ has completely disappeared, and the number of the latter is considerably reduced. The whole is more green and leaf-like than fig. 5.

Fig. 7. The fimbriæ reduced to a single strap; the position of the lateral bundles being indicated by slight projections only. Other instances occurred in which the petal appeared as a single undivided uniform green strap.

Fig. 8. The two exterior whorls of a flower, consisting of seven regularly formed sepals, and eight petals. The latter deviate more or less from the forms represented in fig. 6 and 7. The whole of a green tint, and leaf-like.

Figs. 9, 10. These are parts of one and the same flower dissected to shew the several whorls more distinctly. The whole has assumed a regular appearance, and is composed of seven sepals, alternating with seven green strap-shaped petals, which are succeeded by about twenty stamens without any fleshy disk; the pistil is somewhat metamorphosed. This is perhaps the most remarkable deviation that was noticed from the ordinary state of the flower, and as several examples of it occurred, it is not likely that there is any error in this account of it. It appears to lead us in a very decided manner to the plan on which the flowers of the genus may be considered to be constructed, and to shew us that they are really simple and not compound.

Fig. 11 to 15, represent the appearances assumed by some of the stamens, indicating various degrees of deviation from the perfect state towards a foliaceous structure.

There were other circumstances, besides the appearances in figs. 9. and 10, which may lead us to conclude the structure of the flowers of the genus to be simple and not compound. A compound flower arises from the development of several buds in the axillæ of certain foliaceous appendages more or less degenerated from the character of leaves, and consequently these buds and the florets which they develop are always seated nearer to the axis than the foliaceous appendages themselves. If we suppose a raceme of the mignionette to degenerate into the condition of a compound flower, we must allow for the abortion of the stem on which the several flowers are seated, so that these may become condensed into a capitulum, each floret of which will be accompanied by a bractea, more or less developed, at its base. Let us compare this supposition with the diagrams represented in figs. 16, 17, 18.

Fig. 16. is an imaginary section of the flower in its ordinary state, (a) the pistil, (b) the stamens on the fleshy disk, (c) the petals, (d) the sepals alternating with them.

Fig. 17. represents the position of the several buds (e) which compose the florets of the flower on the supposition of its being compound. Here it will be noticed that these buds alternate with the sepals instead of being placed in their axils where we might rather expect to find them.

Fig. 18. represents a fact which was observed in the present case, where some of the latent buds in the axils of the altered petals were partially developed. This development might perhaps be considered as indicating the construction of a compound flower, and those buds which in ordinary cases compose the outer and abortive florets, it might be said, are here manifesting themselves. But the axes of these buds lie nearer to the axis of the whole flower than the petals in whose axils they are developed; whereas it appears by fig. 17, that they ought to be further from it, since the centres of the five outer circles marked (*e*) would represent the axes of the several buds, whose partial development must be supposed to be on the side next the axis, if we allow any weight to the analogy between the position of the abortive stamens on the supposed calyx, and the fertile stamens on the central disk.

These figures are all that I have thought it necessary to give for the purpose of illustrating the structure of the flower; but as there were several interesting appearances noticed upon dissecting the pistil, I have selected some of them for the second plate, as they may possibly serve to throw some light upon the relationship which the several parts of the ovarium bear to the leaf, and to support the theory of their being all of them merely modifications of that important organ.

Fig. 19. is a pistil in which the three ovules have become foliaceous, and the central, or terminal bud of the flower-stalk is developing in the proliferous form represented in fig. 4.

Fig. 20. The central bud is not developing; but the three axillary buds in the bases of the transformed valves of the pistil are here assuming the form of branches on which one or two pair of leaves are expanded.

Fig. 21. 22. unite the appearances in fig. 19 and 20, with the addition of a glandular body seated between the leaves at their junction. This apparently originates in the union of the two glandular stipules seated at the base of the leaves of this genus, and which may also be seen to accompany the scale-like leaves on the central bud within.

Figs. 23. to 25. Interior views of metamorphosed pistils, in which the ovules are seen transformed to leaves, and the glandular stipules are all that remain of the leaves which should compose the central bud, their limbs having entirely disappeared.

Fig. 26. The appearance of these stipules on a leaf-bud, developing under ordinary circumstances.

Fig. 27. One of them more highly magnified.

Figs. 28. 29. Their appearance on the small scale-like leaves of the central buds in fig. 21, 22.

Fig. 30. Similar to fig. 23, but without any appearance of the transformed ovules; the glandular stipules are seen in the bottom of the ovarium.

These glandular bodies assume a very prominent character in the anatomy of the metamorphosed pistils, and I was for some time puzzled to account for them, thinking that they might represent an altered condition of the ovules. I believe however that I have rightly considered them as the only representatives of the various leaves which would have made their appearance on the branch if the bud had developed in the ordinary way. They do not appear to diminish in size though the limb of the leaf has disappeared.

Fig. 31. Four pedicillated semitransformed ovules, seated on a placenta of a pistil metamorphosed similarly to that in fig. 9.

Figs. 32. to 35. Other appearances of a similar kind, all representing various approaches of the ovules to a foliaceous character. The little theca-shaped appendages are hollow, with a perforation at their apex, representing the foramen.

Fig. 36. One of these dissected, exhibiting a free clavate cellular body within, resembling the columella in the theca of a moss, and here probably representing the nucleus of the ovule.

Fig. 37. In this case the theca-shaped body was partially open exposing the included nucleus.

Fig. 38. This nucleus more highly magnified.

These appearances surely indicate a development of the investing coats of the nucleus into leaves; but how far these developments might be extended, and whether the nucleus itself is capable of being further separated into a series of investing coats does not appear from these specimens.

J. S. HENSLOW.

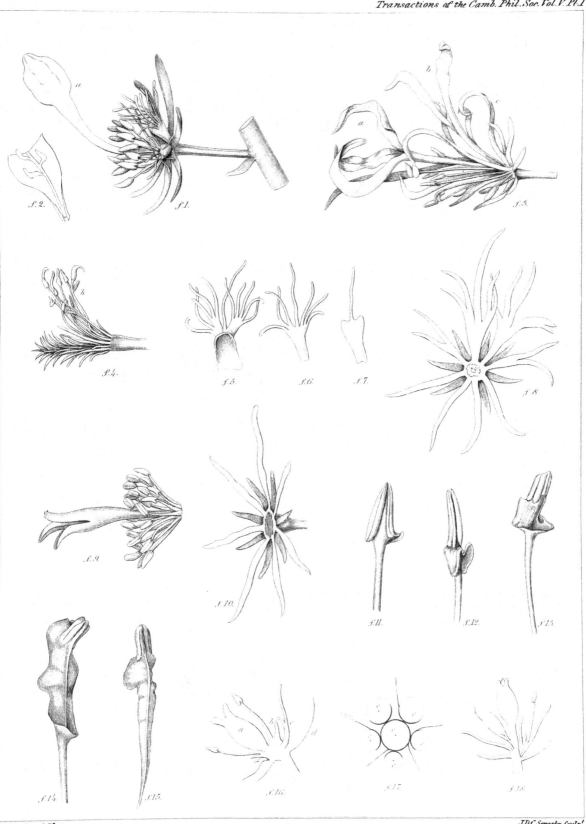

J.S. Henslow del.t

J.D.C. Sowerby sculp.t

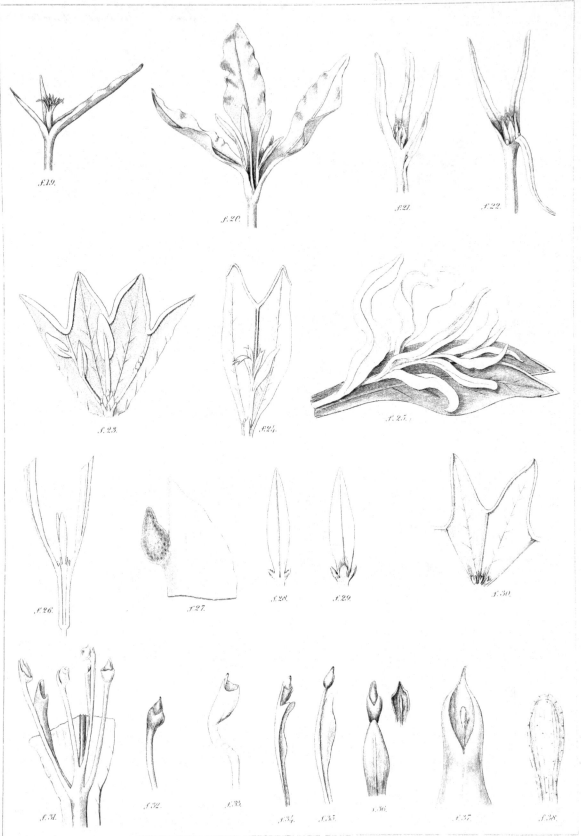

J. S. Henslow del.

J. D. C. Sowerby sculp.

THE

MAGAZINE OF NATURAL HISTORY,

AND

JOURNAL

OF

ZOOLOGY, BOTANY, MINERALOGY, GEOLOGY, AND METEOROLOGY.

VOL. VIII.
1835.

CONDUCTED

By J. C. LOUDON, F.L. G. & Z.S.

MEMBER OF VARIOUS NATURAL HISTORY SOCIETIES ON THE CONTINENT.

LONDON:

PRINTED FOR

LONGMAN, REES, ORME, BROWN, GREEN, AND LONGMAN,
PATERNOSTER-ROW.
1835.

ART. V. *Observations concerning the Indigenousness and Distinct-*
ness of certain Species of Plants included in the British Floras.
By the Rev. J. S. HENSLOW, M.A., Professor of Botany in the
University of Cambridge.

THE present demands of the botanist are not to be satisfied
by receiving a mere description or enumeration of the species
to be found " wild " in any country, but he expects also that
some account will be given of their mode of distribution, and
of various other circumstances of interest in elucidating the
general history of the vegetable kingdom.   The earlier
botanists were often too hasty in admitting certain species
into local floras, upon very unsatisfactory grounds, and later
writers have not always exercised sufficient discrimination in
rejecting them, or, at least, in accurately distinguishing between such as have been introduced by the agency of man,
though now strictly naturalised, and such as are unquestionably indigenous.   Until of late years, many of the authors of
our local floras seem to have been inspired with a desire of
swelling their catalogues, rather than of examining attentively
into the circumstances which may probably, or *possibly*, have
brought certain species within the limits of their observation.
We do, indeed, in many cases, find an (*) attached to some
plants, which were considered as doubtful natives in the district under examination; but when some of these have been
gathered in several distant parts of the country, the original
suspicion of their exotic origin is likely to wear off; and if
any collector, less cautious than others, should be decided in
pronouncing any of them as " truly wild," the chances are
that their claim to rank as " indigenous " species will be
also admitted.   Now, I beg leave to observe, that there is
a wide difference between allowing a plant admission into our
flora as a " wild " and as an " indigenous " species : and,
we may ever hope to arrive at a knowledge of the laws which
regulate the geographical distribution of species, it is of the
highest importance, that all writers of local floras should be
very particular in mentioning the exact circumstances under
which every species in their lists may have been observed to
grow.   Provided this rule be attended to, there can be no
objection to their admitting many more species as " wild
flowers " than they do at present ; and to these they might
append an account of such trees and shrubs as thrive or
fail in plantations, and of such herbaceous plants as succeed
or not in agriculture.   But what I am anxious to impress
upon British botanists is, the propriety of our acting upon
some common principle, in our endeavours to obtain a correct

history of the indigenous vegetation of our country. We ought, at least, to be agreed about rejecting some species which have erroneously crept into our flora, and we ought to endeavour to agree about others which have certainly been introduced by the agency of man. But, besides these, I think we might notice a third class, concerning which I am more particularly desirous of obtaining the opinions of your correspondents and others; to be composed of species, of whose introduction we cannot feel positive that it should be attributed to the agency of man, but of which we have a " suspicion " that they were thus introduced, arising from the peculiar conditions under which they are found. We might indicate this suspicion by some mark, as (†), distinct from that (*) by which we characterise the *certainly* naturalised species. If these dubious marks were to accumulate against the same plant obtained from various stations, we might hunt it, as it were, back to the spot from which it had originally roamed in the society of man to other botanical regions, but in which it would not have been found, had it been left to the operation of natural means alone for its dispersion. The considerations upon which I conceive these marks of doubt may be made to depend, I shall presently revert to.

The idea of soliciting discussion on these points has principally originated in a wish to promote an enquiry which has been ably commenced by Mr. Watson, respecting the geographic distribution of the plants of Great Britain. [See VI. 265. and VII. 20.] This gentleman has rejected many species commonly admitted into our flora, and has increased the list of naturalised species, by including several of those which are usually considered to be strictly indigenous, and has also grouped together, as varieties, several plants which are generally accounted distinct species. These are all points in which our flora stands in need of a decided reformation; and, so far as my own observations have enabled me to judge, I generally agree with Mr. Watson's proposed emendations. We may, possibly, arrive at farther results by a more general discussion of these points, and I know of no place so well adapted to this purpose as the pages of your magazine. I have engaged several of my botanical friends in the University, who are in the habit of meeting once a week for conversing on subjects connected with natural history, to discuss the circumstances under which they may severally have noticed the species included in our flora, so that the communications which I may hereafter transmit to you must be considered as the joint-stock opinions of our resident botanists. There are four heads to which our enquiry will extend : —

1st, To such plants as ought to be entirely excluded from our flora (°).

2dly, To such as have *evidently* been introduced by the agency of man, but are now strictly naturalised (*).

3dly, To such as may be *suspected* of having been originally introduced by the same means (†).

4thly, To those plants usually considered as distinct species, but which there is reason for supposing may be merely varieties.

1. *Plants to be rejected from our Flora.* — It seems hardly correct to include in our lists, even as naturalised species, such as are only occasionally to be met with on heaps of manure, or among rubbish which has been the outcast of a garden. These plants, of acknowledged exotic origin, are for the most part annuals, and seldom occur above once or twice on the spots where they have been observed. The *Datura* Stramònium and *A*marántus *B*lìtum may be instanced as plants of this kind. If any one, however, thinks it desirable that these plants should be noticed in our British flora, they might be placed in an appendix, distinct from the species which are allowed to be indigenous or naturalised.

There are several species which, it is presumed, have been admitted into our flora in mistake for some peculiar varieties of other species, with which they have been confounded. The *P*apàver nudicaúle, and *L*èdum palústre, still retain their places in our lists, though, I believe, they will be omitted in the next edition of the *British Flora,* as Dr. Hooker lately informed me that he was now satisfied that a stunted specimen of Meconópsis cámbrica had been mistaken for the former species, and that the evidence for the latter was equally questionable.

2. *The naturalised Species.* — As it would be certainly improper to exclude those plants from our flora which have become strictly naturalised in our own country, and now form part and parcel of the wild flowers of our fields and hedges, it is perhaps the most convenient mode to register them continuously with the indigenous species, and merely to denote them by the usual mark (*). When a species is considered to be indigenous in one district, but merely naturalised in some other part of the country, this mark may be appended to the names of the places where it has been thus introduced.

There are certain species which are only to be met with in one or two spots of some cultivated districts, and under circumstances where no reasonable doubt can be entertained of their having been originally introduced by agricultural produce. These plants, which are mostly annuals, are continually disappearing and reappearing, according to the nature of the crops which may happen to be under culture at the time, and they are perpetually shifting their quarters from one field to another. Some of these succeed in establishing themselves to a greater or less extent in the hedges, and about the borders of cultivated fields, but are never to be met with in uncultivated districts. Many of these can scarcely be considered as even naturalised, certainly not as truly indigenous plants; but it must be left to the judgment of the observer to decide whether any species in particular is to be placed under this or the following head, until better information respecting its mode of growth in other places has been obtained.

3. *Species possibly not indigenous.* — It is this third enquiry which will afford most scope for discussion; and many persons will at first be disposed to consider the object as of very little importance. But it seems to afford the safest means, at this late period of the earth's history, of arriving at anything like certainty in our conclusions respecting our truly indigenous flora. What has been said under the last head is, perhaps, almost sufficient; but the enquirer may often assist his judgment by obtaining local information, whether any particular species was not formerly cultivated in the neighbourhood, or may not now be growing in a spot which was at one time used as a garden, &c. I once observed *I*ris fœtidíssima and *P*olýgonum Bistórta growing together, in a small patch, in a copse where *H*elléborus fœ tidus abounded, and a few plants of *R*òsa rubiginòsa were also scattered. As these plants were not to be met with elsewhere in the immediate neighbourhood, and as the spot itself was on the outskirts of a village, I was suspicious of their being strictly indigenous; and, upon farther search, detected some straggling shrubs of *B*úxus sempervìrens, which perfectly satisfied me that my suspicions were well founded.

It may be safely asserted of several of our rank weeds, that, unless the ground were cultivated, and the crops regularly grown, they would soon cease to spring up. In fact, although they make their appearance as regularly as the crops themselves, they are seldom to be met with in uncultivated ground; and certainly are not more frequently to be found "truly wild" than individual specimens of the species which compose the crops themselves. When I mention our common field poppies as not exempted from all suspicion of an exotic origin, it will be supposed that I am stating an extreme case; and yet I question whether some if not all the species of the genus *P*apàver (of Decandolle) would not ultimately disappear from our native flora, if the whole kingdom were abandoned to the uncultivated state from which it has been reclaimed for so many generations. I scarcely remember to have seen a specimen of a true *P*apàver in an uncultivated district, unless *P.* Argemòne be not an exception. One good test of the indigenous character of an annual plant is when we see it vigorously disputing possession of the soil with perennials. Thus, the Rhinánthus *Cr*ísta-gálli, which comes up abundantly among the herbage of our pastures, must be considered as truly indigenous; whilst the *Ca*úcalis latifòlia and *d*aucoıdes, which are only to be met with in cultivated fields (in this county, at least), ought not to be considered in any other light than as introduced plants. The *R*anúnculus arvénsis is seldom, if ever, to be met with, except in cultivated districts; but its greater abundance would incline me to place it at present with the poppies, as an indigenous species under " suspicion," rather than as a truly naturalised plant. Some of your correspondents may possibly be inclined to point out other reasons for accepting or rejecting certain species ; and, at all events, they may furnish us with an exact account of the circumstances under which some of the "suspected" species may be found in their respective neighbourhoods. For my own part, I consider it a far more laudable result, to succeed in establishing the exotic origin of a common weed, than to add a new species to our flora, interesting as such an event must be considered by every British botanist.

4. *Plants which may possibly be Varieties, and not distinct Species.* — It is hopeless to expect that botanists will arrive at any safe conclusions on this subject, until they shall consent to adopt the only sure and legitimate mode of satisfying their doubts; I mean, by the test of direct experiment. I have, on former occasions, alluded, in this Magazine [III. 406. 537.; IV. 466.; V. 493.], to the necessity of our adopting this philosophical course, if ever we would hope to arrive at the laws upon which specific distinctions are founded ; and a plan of cooperation was proposed in the botanical section of the British Association, when it met at Cambridge, by which botanists from different parts of the kingdom might be enabled to compare their observations, and obtain more satisfactory conclusions than those at which we have hitherto arrived. It is not too much to say, that there are some genera whose species have possibly been multiplied fourfold beyond the number which they really contain. In consequence of this, our flora appears to occupy a much higher rank among the floras of different countries than it ought to do; and this must lead to very erroneous conclusions respecting the laws which regulate the numerical distribution of species in different latitudes.

MAGAZINE

OF

ZOOLOGY AND BOTANY.

CONDUCTED BY

Sir W. JARDINE, Bart.—P. J. SELBY, Esq.

AND

Dr JOHNSTON.

" Rerum naturalium sagax Indagator."

VOLUME FIRST.

W. H. LIZARS, EDINBURGH;
S. HIGHLEY, 32, FLEET STREET, LONDON; AND
W. CURRY, JUN. & CO., DUBLIN.

MDCCCXXXVII.

VI.—*On the Structure of the Flowers of Adoxa moschatellina.* By the Rev. J. S. HENSLOW, M. A. Professor of Botany in the University of Cambridge.

THE flower of Adoxa moschatellina, as is well known, are arranged in a head, and are so placed that one is terminal, and four others lateral. They are composed of four whorls; but the number of the parts in each is usually different in the terminal and in the lateral flowers. In speaking of the subordinate parts of the two outer whorls we shall consider them as sepals and petals, which cohere to form a gamosepalous calyx and gamopetalous corolla. In this sense then, the terminal flower more usually contains 2 sepals, 4 petals, 8 stamens, and a 4-celled ovarium, which may be considered as compounded of 4 carpels, while the lateral flowers contain 3 sepals, 5 petals, 10 stamens, and 5 carpels (*Fig.* 1.) Such is the ordinary view taken of the structure of these flowers. They are, however, very subject to vary in the number of their parts, and we propose to examine each whorl in detail.

Fig. 1.

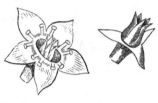

1. *The calyx.* This whorl varies both in the terminal and lateral flowers, with 2, 3, or 4 sepals. In the latter case 3 of the sepals were observed to alternate with 4 petals, and the fourth sepal to be opposite a fifth petal.

De Candolle does not consider this whorl as a true calyx, but calls the corolla a calyx. This reduces the sepals to bracteæ, and as these are combined, the whorl must be considered as an involucrum. In this case we have the tube of the involucrum combining with the lower half of the ovarium, and also uniting with the calyx and stamens. There seems to be no sufficient reason for admitting such an anomaly, and the view usually taken appears to be preferable.

2. *The Corolla.* This is always composed of either 4 or 5 petals, which cohere by their bases and to the upper edge of the calyx, where it becomes free from the ovarium. Some of the petals are sometimes opposite and sometimes alternate with the sepals, which they exceed in number by 1, 2, or 3 parts.

3. *The Stamens.* These are placed very evidently in pairs, a single stamen of each pair standing on either side of the sinus formed

by the union of two contiguous petals. De Candolle asserts that half are opposite, and half alternate, with the petals. This view must be ascribed to a desire to obviate the apparent anomaly of their being neither opposite nor strictly alternate with the petals, but it is decidedly inadmissible. He has not observed that

Fig. 2.

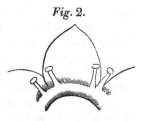

the anthers consist of a single lobe, nor can I discover that this fact has been previously noticed by any author except Dr Hooker, who in his Flora Scotica has the following remark : " Stamens united in pairs, or they may be considered as 4 or 5 forked stamens, each ramification terminated by a single cell of an anther, and all springing from a fleshy ring that surrounds the germen."

I was ignorant of this observation, but was led to make the same remark last spring, by reflecting in what way it would be possible to reduce the anomalous structure of this flower to some normal condition, in which the parts of the several whorls would be arranged agreeably to the

Fig. 3.

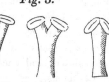

generally established rules of morphology. De Candolle's view appeared to be quite untenable. Upon examining a great number of specimens, I observed in many instances a very decided tendency in the filaments to combine in pairs. Following up the hint which was thus afforded, I found that the anthers were composed of a single cell ; and that it was in those cases only where two contiguous filaments had become completely united, that we ever have a perfect stamen crowned by a two-lobed anther. This at once solved the difficulty, and reduced the structure of the flower within the usual conditions. Dr Hooker has not decided which structure should be adopted, and has followed the arrangement generally adopted of classing this plant under Octandria. The fact of the frequent coherence of the filaments in the contiguous stamens, combined with the regularity thus introduced into the arrangement of the several parts of the contiguous whorls, singularly strengthens the conjecture he had hazarded from the consideration of the anthers being one-celled, and may indeed be considered as affording as direct a proof of the normal condition of this flower as the subject will admit. It may therefore be stated as containing 4 whorls of 5 parts each, the parts alternating in the successive whorls,—a structure eminently characteristic of a dicotyledonous plant, and probably exhibited in greater perfection in the ordinary suppression of 2 or 3 parts in the calyx, sometimes of 1 only; and in the terminal flower in the suppression of one part in all the whorls as they are usually exhibited in the lateral flowers.

Among the numerous specimens which I examined, some had the stamens partially converted to leaves; two small ones being placed facing each other and occupying their usual position on the corolla. (Fig. 4.) In some cases a supernumerary petal of small dimensions was accompanied by a thick filament bearing a three-lobed anther, a monstrosity which apparently originated in a complete and supernumerary stamen having been also developed, and united with the half stamen to which it was contiguous. In other cases, the contiguous filaments were united, and one part foliaceous, the other antheriferous, recalling the structure of the stamens in a Canna.

Fig. 4

Fig. 5.   *Fig. 6.*

Fig. 7.

MAGAZINE

OF

ZOOLOGY AND BOTANY.

CONDUCTED BY

SIR W. JARDINE, BART.—P. J. SELBY, ESQ.

AND

DR JOHNSTON.

" Rerum naturalium sagax Indagator."

VOLUME FIRST.

W. H. LIZARS, EDINBURGH;
S. HIGHLEY, 32, FLEET STREET, LONDON; AND
W. CURRY, JUN. & CO., DUBLIN.

MDCCCXXXVII.

X.—*Description of two new species of Opuntia ; with remarks on the Structure of the Fruit of Rhipsalis.* By Rev. J. S. Henslow, M. A. Professor of Botany in the University of Cambridge.

Sp. 1. *Opuntia Darwinii,*—prostrata, articulis globoso-ovatis, aculearum validioribus elongatis tricuspdiatis, floribus magnis solitariis.
Plate XIV. Fig 1.

The terminal articulation (the only one seen) globoso-ovate, with distant areolæ beset with short tomentum, and those towards the anterior extremity with four to six stiff spines of various lengths, of which the stoutest are one and a half inches long, evidently formed out of three combined, and whose points are free, so that the compound spine appears compressed and tricuspidate. They mostly point forward, but some spread in all directions. Flowers solitary, larger than the articulations which they terminate, yellow. Perianth of six whorls, each of five parts, gradually passing from the form of small fleshy bracteal scales to membranous petaloid segments ; spirally arranged at somewhat more than the fifth of a circle asunder, so as to form five distinct secondary spirals, corresponding to as many, formed by the areolæ on the fleshy tube investing and surmounting the ovarium. These areolæ are placed upon slight tubercular elevations, each bearing a small fleshy bracteal scale, in whose axil is a tuft of yellow tomentum, and those on the upper extremity are also furnished with about half a dozen stiff acicular spines. The segments of the perianth pass gradually from the ovate-apiculate bracteal form of those in the outermost whorl to the cuneato-obcordate, and slightly mucronate petaloid form of those in the innermost, (Fig. *b.*)

Stamens numerous, covering the inner paries of the fleshy tube, (Fig. *c.*) Style remarkably stout, cylindrical, with nine thick radiating stigmata, reaching above the fleshy tube, and a little beyond the uppermost stamens. Ovarium, a small cell, the width of the style, surrounded by the very thick fleshy walls of the lower part of tube or floral receptacle. The character of the herbage appears to agree with that of *Cactus moniliformis,* Lin., which De Candolle places in his division Opuntiaceæ of the genus Cereus; and of which division he says, " An genus proprium inter Cereos et Opuntias medium ?" The flowers of our plant, however, are strictly those of an Opuntia. In assigning the character of " tubum supra ovarium nullum" to Opuntia, De Candolle must consider the whole of the fleshy tubular portion of the receptacle to which the stamens are at-

4

tached as part of the ovarium, which, inceed, it appears to be, when seen from the outside of the flower, but in a transverse section (Fig. *c.*) is evidently prolonged above it.

I have named this interesting Cactus after my friend C. Darwin, Esq. who has recently returned to England, after a five years absence, on board his H. M. S. Beagle, whilst she was employed in surveying the southernmost parts of South America. The specimen figured was gathered in the month of January, at Port Desire, lat. 47°. S. in Patagonia. He recollects also to have seen the same plant in flower as far south as Port St Julian in lat. 49°. S. It is a small species growing close to the ground on arid gravelly plains, at no great distance from the sea. The flowers had one day arrested his attention by the great irritability which their stamens manifested upon his inserting a piece of straw into the tube, when they immediately collapsed round the pistil, and the segments of the perianth soon after closed also. He had intended to procure fresh specimens on the following day, and returned to the ship with the one now figured, but unfortunately she sailed immediately afterwards, and he was prevented from obtaining any more. The geographical position of this species is beyond the limits hitherto assigned to any of the order, which are not recorded as growing much south of the tropic of Capricorn. The climate is remarkably dry and clear, hot in summer, but with sharp frosts during the winter nights. He found Cacti both abundant and of a large size, a little further to the north at Rio-Negro in latitude 41°. S.

Sp. 2. *Opuntia galapageia.*—articulis compressis (saltem junioribus) obovato-rotundis, aculeis setiformibus, longis, penicillatim dispositis, lanugine suprà ovarium densâ.

Plate XIV. Fig. 2.

I am also indebted to Mr Darwin for this second Cactus, of which he brought home in a dry state the specimens here figured. He found it at the Galapagos islands, where it attains to the form of a tree, averaging from six to ten feet high, and about a foot in diameter, perfectly cylindrical or very slightly tapering. The bark is pale-coloured, and in old trees hangs in a ragged manner upon the trunk, which is covered with very strong sharp spines, five to ten of which are set in each fasciculus, in a radiatory manner. From the summit of the trunk numerous branches spread on all sides, somewhat in the manner represented by Fig. *f.*, taken from a very rude sketch of Mr Darwin's. He states these branches to be composed of compressed, rounded, oval articulations, each of which is about one foot in length, wholly without the true spines found on the trunk; but with numerous scattered fasciculi of long elastic bristles strongly resembling hogs' bristles.

Flowers solitary small, like those of *O. Hernandezii,* a few together on the edges of the articulations, (Fig. *b*) red, with the lanugo in the areola on the ovarium dense, especially towards the upper parts, and with a dense tuft also surrounding its base.   Outermost segments of the perianth somewhat scarious, cuneato-obcordate, with an apex, the innermost membranaceous petaloid obovate, emarginate, with a slight mucro. Stamens very numerous and crowded, covering the whole inner surface of the tube.   Style cylindrical, stout, (flexuous ?) with eight stigmata, which are thick, erect, and closely appressed.

This species grows in the lower regions of James's Island, one of the Galapagos, where the soil is extremely arid and rocky, and where it is the only plant of sufficient magnitude to afford any shade, the next in size forming nearly leafless bushes.   The want of water is very great, and the succulent branches of these trees are eagerly devoured by the large tortoises from which the islands have received their name, by large lizards which also abound, and by various other animals. Mr Darwin considers that they are occasionally furnished with a supply of this food from branches broken off by the wind. He found that lizards four feet in length were easily enticed, whenever he threw them a piece of a branch, and small birds would come within a few feet of him and peck at the one extremity, whilst the lizard was eating at the opposite end.

Another Cactus was observed in these islands with the habit of *Cereus peruvianus,* but which does not attain to more than two or three feet in height. It is the first plant that takes possession of the newly formed beds of lava.   Not finding it in flower, Mr Darwin did not preserve a specimen.

Explanation of the Figures.

Fig. I. *(a)* an articulation with a flower; *(b)* a petal; *(c)* a longitudinal section of the flower, shewing the position of the stamens, &c.   All of the natural size.

Fig. II. *(a)* part of a large articulation, with a young one attached to it; *(b)* four flowers, one expanded, and three in bud, seated on the edge of an articulation; *(c)* a longitudinal section of a flower bud enlarged; *(d)* a sepal or outer segment of the perianth; *(e)* a petal or inner segment; *(f)* a rough sketch of a tree.   All, except *c* and *f,* of the natural size.

3

On the Structure of the Fruit of Rhipsalis.

The fruit of Rhipsalis has been considered to possess a different structure from that of all other Cacteæ, in having placentæ in the axis of the berry instead of on the paries, and the berry itself has also been considered as probably trilocular, whilst those of all other genera in the order are unilocular. Hence De Candolle has placed it in a separate tribe, his Rhipsalideæ, expressing, however, some doubts about the accuracy of the observations upon which he has founded his arrangement. In his last memoir on the Cacteæ (1834,) he separates *Rhipsalis salicornioides* from the rest, under the generic name of Hariota ; and in that species he states that he had ascertained the ovary to be unilocular, and the placentæ parietal. I have lately had an opportunity of examining the fruit of *Rhipsalis cassytha* in all stages of its growth, and can safely assert that both suppositions, of its being trilocular and having central placentæ, have originated in a mistake. When the fruit is ripe, the seeds are nestled in the midst of a very liquid pulp, and are no longer attached to any part ; but in earlier stages of its growth, they are found to adhere in double rows upon three placentæ, disposed longitudinally on the paries. At first sight there is a deceptive appearance of three dissepiments, or at least of three inwardly projecting placentæ, to the innermost extremities of which the seeds are attached ; but further examination shews this to arise, merely from the close agglomeration of the funicular chords (see Fig. A,) which stretch from the paries towards the axis, and from whose extremities the ovules are suspended in a reversed position. The placentæ themselves scarcely form any projection on the paries, as is very evident in some cases where many of the ovules have become abortive ;

A

and, indeed, several are so in all cases, and then appear as small brown spots attached to the paries. It seems to me likely that the watery pulp in which the seeds are nestled in this and other genera of the order is derived from the super-developement of the cellular tissue of the funicular chords. The whole coat of the berry, formed by the union of the calyx tube and pericarp, is very succulent, but the interior pulp is much more so ; and this does not appear any way connected with, or to originate from the inner coats of the pericarp, but in the way here suggested. Be this as it may, it is sufficiently evident that the ovary and fruit of Rhipsalis are strictly unilocular with parietal placentæ, as in all other Cacteæ, and, consequently, it is necessary that the tribe Rhipsalideæ should be suppressed.

ANNALS OF NATURAL HISTORY;

OR,

MAGAZINE

OF

ZOOLOGY, BOTANY, AND GEOLOGY.

(BEING A CONTINUATION OF THE 'MAGAZINE OF ZOOLOGY AND BOTANY,' AND
SIR W. J. HOOKER'S 'BOTANICAL COMPANION.')

CONDUCTED BY

Sir W. JARDINE, Bart.—P. J. SELBY, Esq.,
Dr. JOHNSTON,
Sir W. J. HOOKER, Regius Professor of Botany,
AND
RICHARD TAYLOR, F.L.S.

VOL. I.

LONDON:
PRINTED AND PUBLISHED BY R. AND J. E. TAYLOR.
SOLD BY S. HIGHLEY; SIMPKIN AND MARSHALL; SHERWOOD AND CO.; W. WOOD,
TAVISTOCK STREET; BAILLIERE, REGENT STREET, AND PARIS:
LIZARS, AND MACLACHLAN AND STEWART, EDINBURGH:
CURRY, DUBLIN: AND ASHER, BERLIN.
1838.

ANNALS OF NATURAL HISTORY.

XXXVII.—*Florula Keelingensis. An Account of the Native Plants of the Keeling Islands.* By the Rev. J. S. HENSLOW, M.A., Professor of Botany in the University of Cambridge.

THE Keelings consist of small coral islands, ranging in a circle, and inclosing a lagoon or salt-water lake of nine and a half miles in its longest diameter. They lie in lat. 12° 5′ S., and long. 90° 55′ E., very nearly 600 geographical miles to the S.W. of Java Head or the Straits of Sunda. They stand apart from any other group or archipelago, and the naturalist is curious to learn the character of their productions. Mr. Darwin, who accompanied the Beagle in her late voyage round the world, visited these islands in 1836, and is about to give an account of their geological conditions, as well as of the scanty zoology which they furnish. As he obligingly presented me with the plants which he collected, together with his memoranda respecting them, I have thought that a list of the species, accompanied by a few remarks, might be of interest; and chiefly as serving to point out a set of plants whose seeds must be provided in a very eminent degree with the means of resisting the influence of sea water. For the satisfactory determination of the geographical distribution of species, it is necessary to be extremely careful in discriminating the species and even varieties which occur in different regions, and I have therefore generally added a few remarks on the state of the individual specimens in question, that every one may form a better estimate of the degree of probability of each having been correctly identified.

The largest of the islands is about five miles long and a quarter of a mile broad. Some sand hillocks on it are thirty feet in height, but the general level does not exceed six or eight feet. The foundation of all of them is a solid coral reef, which receives continued additions from fragments of coral and sand brought by the waves and wind. The soil is entirely composed of broken corals and shells, sometimes in the form of calcareous sand; and the quantity of vegetable mould is extremely small. Twenty-three of the islands bear trees; and there are many others of small dimensions, scarcely elevated above the surface of the ocean, which produce none. When first seen, nothing can be observed but a belt of cocoa-nut trees encircling the lagoon. The abundance in which these occur has tempted a respectable Englishman named Ross to bring his family and settle here. He has with him a party of about eighty Malays, who are employed in manufacturing cocoa-nut oil; and the nuts also are exported to Mauritius and Singapore. Thrown as these men are so completely upon their own resources, they have accurately investigated the natural productions of the islands, and readily pointed out to Mr. Darwin the different species of plants, and assured him that he had seen them all except one, of which there was only a single tree, bearing a large square and very hard nut, growing on one of the islands which he did not visit. Excepting the cocoa-nut, and one other tree which was not in flower, and which attains a diameter of five or six feet, with particularly soft wood, Mr. Darwin brought away specimens of all the species he saw, amounting to twenty-one.

From the character of the soil and the condition of the islands we might expect à *priori* to meet with a purely littoral flora, and with none but extensively sporadic species. Mr. Darwin heard of the trunks of trees, of many seeds, and of old cocoa-nuts being washed on shore from time to time, and probably all the species which have thus been introduced are to be found in the East Indian Archipelago, or on the neighbouring continent, though they have not all been noticed there. Two at least of the species appear to be hitherto undescribed, and one or two others possess an interest from their rarity, and the little information we possess concerning them; but all the rest have an extensive range throughout the intra-tropical regions.

Of the few imported plants the banana does not thrive well; the sugar cane has in some parts run wild, but has lost greatly in flavour, as also has the tobacco. Besides these a little maize and a few vegetables are cultivated. Three species of grass had been introduced, (*Panicum . . . . . . , Eleusine indica,* and *Poa plumosa,*) as was stated, from Java, under an impression that goats would not eat the rank herbage of the island; but the settlers were surprised to find that one of these animals left on the islands by Capt. Fitzroy preferred the native to the imported species.

As the flora of the island of Timor, which lies nearly due west of the Keelings without any intervening land, has lately been described by Mons. Decaisne, I have placed a (T) in the following list opposite those species which he has recorded in his very excellent ' Herbarium Timorense.'

List of the Plants Indigenous to the Keelings.

MALVACEÆ.
1. Paritium tiliaceum, *St. Hil.* T.
TILIACEÆ.
2. Triumfetta procumbens, *Forst.*
LYTHRACEÆ.
3. Pemphis acidula, *Forst.* T.
PORTULACACEÆ.
4. Portulaca oleracea.
LEGUMINOSÆ.
5. Guilandina Bonduc, *Hort. Kew.* T.
6. Acacia (Farnesiana?) *Linn.* T.
URTICACEÆ.
7. Urera Gaudichaudiana, *n. s.*
AMARANTHACEÆ.
8. Achyranthes argentea (var.?) *Lam.* T.
NYCTAGINACEÆ.
9. Boerhavia diffusa, *Willd.* T.
var. β.?
var. γ.?
SCÆVOLACEÆ.
10. Scævola Kœnigii, *Vahl.* T.

CINCHONACEÆ.
11. Guettarda speciosa, *Linn.* T.
CORDIACEÆ.
12. Cordia orientalis, *R. Brown.* T.
BORAGINACEÆ?
13. Tournefortia argentea, *Linn.* T.
ACANTHACEÆ.
14. Dicliptera Burmanni, (var.?) *Nees.*
APOCYNACEÆ.
15. Ochrosia parviflora.
GRAMINEÆ.
16. Panicum sanguinale, (var.?) *Linn.* T.
17. Stenotaphrum lepturoide, *n. s.*
18. Lepturus repens, *Forst.*
PALMÆ.
19. Cocos nucifera, *Linn.* T.
MUSCI.
20. Hypnum rufescens, *Hooker.*
FUNGI.
21. Polyporus lucidus.
22. } Two trees of which no spe-
23. } cimens were procured.

1. *Paritium tiliaceum.*—Leaves large, and the linear pore upon one to five of the nerves on the under side.

" Common on one of the islands. It is exceedingly useful throughout the Pacific; and in Otaheite particularly, the bark is employed in the manufacture of cordage, whilst the light wood is used by the fishermen for floats. The natives readily procure fire from the wood by friction."—*C. Darwin.*

2. *Triumfetta procumbens.* Forster, Prod. n. 204.—This species is placed by De Candolle among those " non satis notæ." By Mr. Brown's kindness I have satisfactorily iden-

tified it, by comparison with Forster's original specimens in the British Museum. As much uncertainty prevails respecting the number of species in the genus, I shall add a detailed description of the present specimens. Messrs. Wight and Arnott have observed, at page 74 of their Prod. Floræ Indiæ : " In this genus it may be right to caution the student to place no reliance on the shape of the leaves or their pubescence, or suppression of the parts of the flower." To this we would add further, that neither can much reliance be placed upon the character of the inflorescence, since the differences between the peduncles being axillary or opposite, seem chiefly to depend upon different degrees of luxuriance.

> Speciminum Keelingensium caules ramosi, ramis tomentosis, pubescentiâ stellatâ. Folia longè petiolata, subrotunda vel latè-ovata, cordata, indivisa vel trilobata, inæqualiter serrato-crenata : suprà nudiuscula, subtus petiolisque incano-tomentosa, marginibus nudis subglandulosis. Stipulæ lanceolatæ. Pedunculi inferiores axillares, sub-abortivi; superiores oppositifolii, foliorumque superiorum abortione sub-corymbosi, horumque stipulis bracteas emulantibus; pedicellis 3—5 sub-umbellatim dispositis. Calyx, sepalis 5 linearibus, sub apice acuminatis, extus pubescentibus, æstivatione valvatis. Corolla, petalis 5, sepalis parum minoribus, obovatis, unguibus basi villosis. Stamina 25, petalorum longitudine. Pistillum ovario ovali, hispido; stylo lineari, hirto, tricuspidato. Capsula junior globosa uncinato-hispida.

3. *Pemphis acidula.*—The capsules burst by an irregular transverse fissure about the middle, with the lower portion more membranous than the upper. Forster describes them as having six valves, and Lamarck as opening transversely at the base.

" No sooner has a new reef become sufficiently elevated by the accumulation of sand upon its surface, but this plant is sure to be the first which takes possession of the soil."— *C. Darwin.*

4. *Portulaca oleracea.*—The specimen is in seed, tolerably luxuriant, and seems unquestionably to belong to this species; but there are some minute hairs in the axils, which is not generally the case, and not characteristic of the section to which it belongs.

5. *Guilandina Bonduc.*—The specimen is only in bud.

" Grows only on one islet."—*C. Darwin.*

9   V   K V   K V

10

11

Achyr: argentea.

2   3   1

4   5   6

Urera Gaudichaudiana.

6. *Acacia* (*Farnesiana* ?)—The specimen has no signs of inflorescence, but the herbage closely resembles that of *Farnesiana*; and as that species grows in Timor, it is probably the same.

" On the same islet with the last."—*C. Darwin.*

7. *Urera Gaudichaudiana.* Plate XI.

Caule herbaceo; foliis longè petiolatis, latè cordatis, sub-acuminatis, grossè serratis, undique pilis brevibus conspersis, subtus pallidioribus; cymis axillaribus divaricato-dichotomis petiolis subæqualibus.

I have named this species in honour of Mons. Gaudichaud the founder of the genus *Urera,* who has attempted to group the species of this much-neglected order in the volume devoted to the botany of the ' Voyage de l'Uranie.' The only described species to which it seems to approach is the *Urtica ruderalis* of Forster, but a comparison with his original specimen in the British Museum has shown me that it is perfectly distinct.

The single specimen brought home by Mr. Darwin consists of part of an herbaceous stem about seven inches long, belonging apparently to a perennial. From each of the axils of the two lowermost leaves proceeds a short branch, and from each of seven or eight others spring divaricate branching cymose panicles about four inches long. The petiole and limb of the largest leaf are each four inches long, and the latter is 2½ inches broad. The inferior panicles produce male flowers on their lower branches and female on their upper; but the superior bear female only. Male flowers crowded in small heads at the extremities of the short branches, their calyx deeply 5-partite (fig. 1.); stamens 5. Female flowers smaller than the males, their calyx of three sepals, or rather of two sepals and an external bract (fig. 2.); the pistil solitary, ovary ovate and slightly oblique (fig. 3.); the stigma crowned with a ferruginous tuft of hair inclining to one side. The ripe pericarp obliquely-ovate or gibbous (fig. 4.) containing one erect sessile exalbuminous seed (fig. 5.) with the embryo inverted (fig. 6).

8. *Achyranthes argentea* (var. ? *villosior.*)

Foliis breviter pedicillatis, oblongis, basi sub-attenuatis, superne villosis, subtus incano-scriceis.

There are two specimens of this, each about a foot long, with the terminal spike on one of them six inches, on the other not two. Largest leaves three inches. It is difficult to decide whether this ought to be considered a new species or only a variety of *argentea*.

Decaisne considers *argentea* and *aspera* to be identical. The very variable character of the herbage prevents our laying any great stress upon the shape of the leaf, length of the spike, or degree of pubescence. In these respects our plant comes within the character of *argentea* given by Decaisne in the 'Flora Timorensis.'

On comparing the several parts of the flower with those of another specimen of *argentea*, brought by Mr. Darwin from the Cape-de-Verd Islands, I find several remarkable differences, which I may here describe.

Comparison of the parts of the flower in specimens of Achyranthes argentea *from the Keelings and Cape-de-Verd.   Plate XI. where K. means Keeling, and V. Cape-de-Verd Islands.*

KEELING.	CAPE-DE-VERD.
Fig. 7. *Bract.* Auricles at base, about half the length of the bract.	About one third the length.
8. *Sepal.*	
9. *Stamens and pistil.*	
10. *Stamen, with part of connecting membrane.*	
Anther. Elliptic-oblong, equal to free portion of filament.	Subrotund and much shorter.
Fringed lobes (from abortive stamens?) with few and regular incisions.	Incisions numerous and very irregular.
11. *Pistil.* Ovary ob-ovato-globose, depressed, with the style three times as long.	Ob-ovato-cylindrical, with the style half as long.

The position and form of the ovule is also marked on the figures.

9. *Boerhavia diffusa.*—After an attentive examination of Mr. Darwin's specimens, I cannot detect sufficient differences to class them under more than one species, though he had himself concluded, from certain peculiarities in their habit whilst growing, that they must belong to three. These three forms, which I consider to be varieties of the *diffusa* of Decaisne's Herb. Timor., have each long, weak, straggling, terete branches, clothed with close scattered pubescence, except on the older parts, which are glabrous. The leaves are stalked

and fleshy, modifications of ovate and repand. The flowers in small heads, which themselves are arranged in dense umbels, with long axillary peduncles alternately disposed among the uppermost parts of the branches.

Var. a. Stoutest in habit, and with the largest leaves, the lowermost of which have their limb an inch long, with peduncles of half an inch; all are pedunculate, ovato-rotund, often slightly sub-cordate, much paler beneath. Stamens 2—3; young fruit ob-clavato-fusiform.

Var. β. Branches more than three feet long. Leaves rather smaller and darker on each side, generally more acute, the uppermost nearly or quite sessile. Seems to be *B. diandra* of Bur. Fl. Ind., tab. 1. fig. 1. Stamens 2—4, alternate with the segments of the calyx; anther with two globose cells, which, with the filaments, are pilose. Ovary oval, but in the young fruit becomes fusiform and angular, with glandular hairs. Stigma peltate. A toothed annulus round the calyx was noticed in one specimen. Three or four bracts.

" Grows upright and untidy, and is the commonest weed, growing everywhere."—*C. Darwin.*

Var. γ. Branches a foot and a half long. More stunted, with fewer, smaller, and more fleshy leaves. Stamens 2—3.

" Grows close to the ground, and is abundant on one spot within ten or twelve yards of the sea, where it was pointed out to me as possessing an esculent root, and considered to be quite distinct from var. β."—*C. Darwin.*

A specimen of the root was preserved, and consists of long wiry branches, which do not appear to have been ever very succulent.

10. *Scævola Kœnigii.*—The leaves are seven inches long and three broad, quite glabrous ; the apex slightly retuse and the margin somewhat repand. Segments of the calyx subulate and glabrous. Corolla with the base of the tube slightly villose within, the segments of the limb lanceolate and glabrous. Cupula of the stigma very pilose within. This specimen appears to be more glabrous than usual, whilst *S. sericea* (of which I have specimens from Macao in China) differs from the more usual state of *S. Kœnigii* chiefly in being more decidedly pubescent.

11. *Guettarda speciosa.*—Largest leaves eleven inches long and nine broad. Corolla with seven or eight segments. Stamens 7—8. Ovary seven cells with a pendulous ovary in each. Stigma eight rays. Pollen intermixed with numerous fibres (pollen tubes?).

" The flowers possess a delightful perfume."—*C. Darwin.*

12. *Cordia orientalis.*—" The settlers have named this Keeling-teak, because it furnishes them with excellent timber. They have built themselves a vessel with it. A large tree, abounding in some of the islands, very leafy, with scarlet flowers; but only a few blossoms were expanded at the time, and they easily fell off."—*C. Darwin.*

13. *Tournefortia argentea.*—Cyme ten inches long, bearing both flower and fruit. Leaves oblong and obovate-oblong, attenuated below.

" A moderate sized tree, with small white flowers, very common."—*C. Darwin.*

14. *Dicliptera Burmanni*, var. ?—Some of Nees von Esenbeck's species (in Wallich's Pl. As. Rar. vol. iii. pp. 111, 112,) run so closely together, that it is difficult to say whether he would have referred these specimens to *Burmanni* or not. I will here subjoin a full description of them, and it may serve future observers in either extending the character of *Burmanni*, or of reuniting with it some of the other forms now considered to be distinct species, but formerly combined under the name of *Justicia chinensis.*

Radix annua ramosa. Caulis obsoletè tetragonus. Folia inferiora 4 pollices longa, 2¼ lata, petiolo unciali, subglabra strigosave, subtùs pallidiora, cum caule lineolata; foliorum margines pilis minutis appressis tectæ, et basim versus aliquando piloso-ciliatæ. Axillæ plerumque floriferæ. Pedunculi 4—6 in quâve axillâ seriatim dispositi, 1—2 lineares, majores interiores. Capitula 1—2-flora. Bracteæ primariæ (sive umbellarum) plerumque subulato-cuspidatæ, pungentes, 6-lineares; aliquando inter umbellas inferiores eâdem secundariarum formâ, sed majores et foliaceæ. Bracteæ secundariæ (sive capitulorum) vel subspathulatæ vel obovatæ vel lanceolatæ vel lato-ovatæ, basi pallidiori attenuato, nervo medio valido, in apicem cuspidato-mucronatum excurrente, hirsutæ, pilis longis articulatis glandulisque interjectis ciliatæ. Bracteæ tertiariæ (sive florum bracteolæ) binæ setaceæ, calyce sublongiores. Calyx subsessilis minutus 5-partitus, laciniis subsetaceis, bracteolisque hirsutæ et ciliatæ. Corolla 7-linearis, tubo pallido, limbo roseo bila-

biato, labio superiore breviter 3-dentato, inferiore obsoletissime 2-den-
tato, externe pubescens. Capsula orbicularis, tomentosa, compressa
ungue brevi dorsaliter compresso. Semina duo, orbicularia, compressa,
muricata, primum pallide denique autem saturatissime brunnea.

15. *Ochrosia parviflora.*—This is unquestionably the *Cerbera parviflora* of Forster Prod. n. 121., as Mr. Brown showed
me by comparison with the original specimens in the British Museum; but Dr. Hooker's *C. parviflora*, in Beechy's
Voyage, p. 90, is certainly a distinct species, as I have ascertained by an examination of his specimens, kindly forwarded
to me for comparison with Mr. Darwin's. Dryander, in
the Linn. Trans., vol. ii. p. 227, asserts that he had compared
Forster's specimens of *C. parviflora* with Commerson's of
Ochrosia borbonica, and found them to be the same species.
This has been since disputed. I have specimens of *Och. un-
dulata* from Mauritius, labelled by Bojer as the " *Bois jaune*"
of that island, which appears to identify that species with
Jussieu's *Och. borbonica.* There is some obscurity in the descriptions hitherto given of the fruits of *Cerbera, Ochrosia*
and *Tanghinia,* and I had hoped to have been able to have
inserted here my own observations on them, but I must defer
them until I have time to clear up one or two points about
which I am doubtful. I should feel much obliged in the
mean time to any botanist who can furnish me with specimens
of the fruit of these, or any allied genera, for dissection. Mr.
Darwin's specimens were accompanied by the following note :
" Forms straight handsome trees, with smooth bark, which
are commonly dispersed two or three together. The fruit is
bright green, like that of the walnut." Two specimens of this
fruit were brought home, and though Mr. Darwin feels confident that he *gathered* them, and, as he believes, from the
same tree which bore flowers at the time, yet it has been supposed that they must belong to a species of *Cerbera,* and not
to an *Ochrosia* which this plant seems to be ; and I shall therefore defer their description for the present, merely intimating
that I believe them to be identical with the *Cerbera platy-
sperma* of Gærtner. The following is a detailed description
of the flowering specimens from Keeling.

Folia subternata (quorum longiora cùm petiolo sesquipedalia, limboque
decem pollices longo sex lato), oblonga vel obovato-oblonga, subacumi-

nata, basi parum attenuatâ, subcoriacea, integerrima, glabra, subtus pallidiora, venis secundariis transversis parallelis, numerosis, supernè incurvantibus. Pedunculi axillares vel foliorum abortione extra-axillares et terminales, sub-ternatim verticillati, petiolis longiores supernè di-tri-chotomi. Flores breviter pedicellati bracteis binis suffulti, densè corymbosimque dispositi. Calyx parvus 5-partitus. Corolla semiuncialis fauce parum inflatâ, limbo quinquifido. Stamina 5, antheris acuto-ovatis, filamentis brevibus. Pistillum è carpellis duobus in ovarium biloculare primùm accretis, subitò in drupellas duas sex ovulatas segregans; ovulis 2—4 solummodo maturascentibus ?

16. *Panicum sanguinale*, var. ?.—I quote this species with doubt, because the only specimen has the spikes half starved and the spikelets not fully matured. It has much the habit of *P. pruriens* of Trinius Gram. Icones, with a trailing stem of four feet, but the glumes have the relative proportions ascribed to *P. sanguinale*, and the margins of the superior one are very hirsute. There are thirteen spikelets, but three or four towards the summit are quite abortive. They are arranged in two whorls of four in each; one is below the lowest whorl, and the other four are scattered between the two whorls. As Decaisne gives *P. sanguinale* as a Timor plant, the present may the more probably be only a form of this.

17. *Stenotaphrum lepturoide.* Plate XII.

Spiculis subduabus alternatim dispositis, unâ rachi sessili, alterâ pedunculatâ, foliis lanceolatis lineari-lanceolatisque.

Mr. Brown showed me a single specimen of this grass among Forster's specimens of *Lepturus repens* in the British Museum, and the general resemblance which it may be considered to bear to that plant has induced me to give it the specific name of *lepturoide*. It departs from the generic character of *Stenotaphrum*, given in Kunth's Agrostographia, in not having the spikelets arranged unilaterally, and in the rachis of the spike being terete or very nearly so; but in all essential points it is truly a *Stenotaphrum*, as the following detailed description will be sufficient to show.

Culmi pedales et ultrà, ramosi, procumbentes vel supernè solummodo ascendentes, plerumque fertiles, glabri, compressi, nodis brunneis. Folia lanceolata vel lineari-lanceolata, acuta, plana, nervis 9 subprominulioribus, intermedio subtus validiore, membranaceo-rigida, utrinque glabra, marginibus obsoletè scabriusculis, 1—2 poll. longa, $1\frac{1}{2}$—3 lin. lata. Vaginæ ad basim fissæ, marginibus primùm ciliato-pilosis, ore pilosiore, unipollicares, plerumque solutæ. Ligulæ obsoletæ, vel in lacinias breves resolutæ. Spicæ in apice ramorum solitariæ, basi è

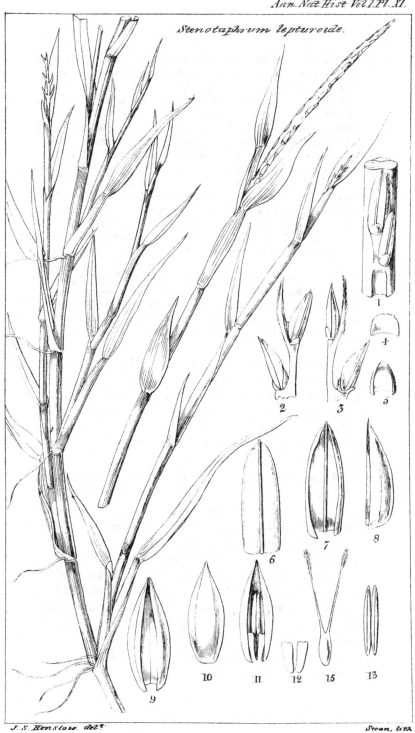

Stenotaphrum lepturoide.

1
2 3 4 5
6 7 8
9 10 11 12 15 13

J. S. Henslow, delt.   Swan, lith.

summâ vaginâ exsertâ, subincurvantes, 3—4 poll. longæ; rachi tereti,
vel paululum compressâ (fig. 1.) acutâ, internè spongiosâ, vix lineam
latâ, pro insertione spicularum utrinque excavatâ. Spiculæ per binas
(fig. 1.) (vel inferne per ternas), quarum una sessilis, altera pedicellata
(figs. 2, 3.), ovato-oblongæ, lineæ dimidio longiores, bifloræ, flore in-
feriori unipaleaceo (figs. 6, 7.) neutro; superiori hermaphrodito (fig. 8.)
bipaleaceo. Glumæ duæ subæquales (figs. 4, 5.) concavæ, enerviæ,
membranaceæ, glabræ, spiculâ quadruplò breviores, ovato-ellipticæ, ex-
terioris (fig. 4.) apice sub-truncato eroso. Flos neuter è paleâ unicâ
ovato-ellipticâ dorso planâ (fig. 6.), nervis 3 prominulis, medio sub-
carinante excurrente acutâ, glaber, coriaceus, florem hermaphroditum
unilateraliter amplectens et paululum superans. Flos hermaphroditus
(fig. 8.) ovato-oblongus, sub-acuminatus, externe convexus, interne pla-
niusculus, pallidus, lævis, glaber, paleis duobus, quarum inferior (fig. 9.)
oblongo-ovata, acuta, concava, trinervis, superiorem amplectens, mem-
branaceo-chartacea; superior (figs. 10, 11.) ovata, binervis, concava,
dorso (fig. 10.) planiuscula, marginibus inferne inflexis. Squamulæ
(lodiculæ) (fig. 12.) duæ anticæ, collaterales, truncato-lineares, ovario
longiores. Stamina 3, antheris (fig. 13.) lineari-oblongis. Ovarium
(fig. 15.) oblongum, apice in stylos duos elongatos attenuatum. Stigmata stylis duplò breviora, plumosa, pilis brevioribus, simplicibus, hya-
linis.

18. *Lepturus repens.*—" Occurs in salt places, in the interior of the islands."—*C. Darwin.*

19. *Cocos nucifera.*—Although no specimen of this was
brought home, yet as the Keelings are also called Cocos
Islands, and as they have been recently colonized for the ex-
press purpose of trading in the oil and fruit, we may safely
assert it to be abundant.

20. *Hypnum rufescens.*—The specimens were submitted to
Dr. Hooker, who remarks, " In a very indifferent state certainly, but I think it may safely be referred to *H. rufescens*,
Hooker and Arnott, of Bot. of Beechy's Voyage, page 76, t.
19. It is in a younger and greener state."

21. *Polyporus lucidus.*—These were sent to Mr. Berkeley
with a query, whether they might not be *P. australis*; to which
he replies, " I have no doubt your fungus is *P. lucidus*. I have
before me specimens of precisely the same thing from Mauritius, together with a distinct variety resembling, I should
imagine, *P. australis*. That, however, is a perennial species,
and the substance is very hard; whereas your plant is at most
biennial, and the substance soft and spongy."

Printed in the United States
By Bookmasters